MW01716288

AMERICA AND THE INDO-PACIFIC

This book offers an extensive account of Donald Trump's foreign policy record in the Indo-Pacific region. Set against the backdrop of Trump's policy of sustained US confrontation with China, it recounts his administration's efforts to shore up America's position with the Indo-Pacific strategy. It also reviews Trump's record with allies and partners in the Indo-Pacific and the South Asian subregion in context of the 'great power competition' between China and the United States. Amidst the ongoing conversations on the declining currency of American internationalism, the volume showcases the seeming insularity of the Indo-Pacific region from forces that are informing an America in retreat. In noting Trump's record to have been a consequential one, the authors also offer insights into the prospects for US policy continuity under Joe Biden.

This timely book will be of great interest to scholars, teachers, and students of politics and international relations, Asia studies, US–China studies, area studies, foreign policy, maritime studies, and world politics. It is a recommended read for all watchers of US foreign policy and the evolving US–China rivalry.

Harsh V Pant is Professor of International Relations at King's College London. He is Director, Studies and Head of the Strategic Studies Programme at Observer Research Foundation, New Delhi, India. He is also Director (Honorary) of Delhi School of Transnational Affairs at Delhi University. His current research is focused on Asian security issues. His most recent books include *New Directions in India's Foreign Policy: Theory and Praxis* (2018), *Indian Nuclear Policy* (2018), *The US Pivot and Indian Foreign Policy: Asia's Evolving Balance of Power* (2015), *Handbook of Indian Defence Policy* (2015), and *India's Afghan Muddle* (2013).

Kashish Parpiani is Fellow at Observer Research Foundation, Mumbai, India. His research interests include US–India bilateral ties and US foreign policy in the Indo-Pacific. He has published with the *Institute for National Security Studies* (Israel), the *Centre for Asia and Globalization* (Singapore), *Deccan Herald* (India), the *Marine Corps University Press* (US), *The Hindu* (India), and the *Rising Powers in Global Governance Project* (Turkey). He has presented papers at conferences organised by the Shanghai University, the International Studies Association (ISA), and London School of Economics–IDEAS. On varied matters pertaining to US–India ties and US foreign policy, he has been quoted in media platforms such as *The Wall Street Journal*, *South China Morning Post*, *Foreign Policy*, *Bloomberg*, *TIME* magazine, and *Al Jazeera*.

AMERICA AND THE INDO-PACIFIC

Trump and Beyond

Harsh V Pant and
Kashish Parpiani

Routledge
Taylor & Francis Group
LONDON AND NEW YORK

First published 2022
by Routledge
2 Park Square, Milton Park, Abingdon, Oxon OX14 4RN

and by Routledge
605 Third Avenue, New York, NY 10158

Routledge is an imprint of the Taylor & Francis Group, an informa business

© 2022 Harsh V Pant and Kashish Parpiani

The right of Harsh V Pant and Kashish Parpiani to be identified as authors of this work has been asserted by them in accordance with sections 77 and 78 of the Copyright, Designs and Patents Act 1988.

All rights reserved. No part of this book may be reprinted or reproduced or utilised in any form or by any electronic, mechanical, or other means, now known or hereafter invented, including photocopying and recording, or in any information storage or retrieval system, without permission in writing from the publishers.

Trademark notice: Product or corporate names may be trademarks or registered trademarks, and are used only for identification and explanation without intent to infringe.

British Library Cataloguing-in-Publication Data
A catalogue record for this book is available from the British Library

Library of Congress Cataloging-in-Publication Data
A catalog record for this book has been requested

ISBN: 978-1-138-57070-2 (hbk)
ISBN: 978-1-032-05105-5 (pbk)
ISBN: 978-1-003-19603-7 (ebk)

Typeset in Sabon
by Apex CoVantage, LLC

CONTENTS

PREFACE

The idea of this book germinated in the early years of the Trump presidency when the widespread structural, institutional and normative disruption was the order of the day. At that time, we decided to embark on a book project that was aimed at exploring Trump's approach to Asia. But for various reasons, the project got delayed and by the time we started to work again, the four years of President Trump were coming to an end. America's approach to the Indo-Pacific had taken a much more concrete shape and the challenge of the COVID-19 pandemic had ensured that most of the old assumptions about the international system had to be re-examined.

This book presents a detailed account of Donald Trump's foreign policy record in the Indo-Pacific region and explicates his administration's efforts to shore up America's position with the Indo-Pacific strategy. It examines Trump's engagements with America's allies and partners in the Indo-Pacific, amidst the wider shift in the regional balance of power and a sharpening contestation between the US and China. In more ways than one, this book underscores the seeming insularity of the Indo-Pacific region from forces that are informing an America in retreat.

We would like to thank Shashank Shekhar Sinha of Routledge India who first pushed us into taking up this project. Though we failed to meet his initial time-line, we do hope that this volume has justified his efforts to invest in this endeavour. We are grateful to Shashank and his entire team for bearing with us and having the faith in our ability to deliver when at times we ourselves faltered. We would also like to thank the Observer Research Foundation for giving us the much needed time and space to complete this project. And finally, thanks to our families for all their support!

INTRODUCTION

Donald Trump's election victory in 2016 put the future of American internationalism in a state of limbo. On the campaign trail, Trump purported a conservative nationalist vision for US foreign policy – which sought to upend post–Cold War American advocacy of globalisation, furtherance of liberal *Wilsonian* values, and sustenance of America's constellation of allied and partner nations (some of which even date back to the end of the Second World War). Apprehensions over Trump's America First proposition – which fast devolved from merely being subject to allegations of isolationism to having spurred the untoward reality of unilateralism at its worst in terms of 'America alone', were compounded by the emergence of new age threats to the international system.

Away from yesteryears' state-led or state-centred threats, by 2016, the world had witnessed newer challenges, such as the diffusion of terror networks with lone-wolf attacks, rise of multipolarity further problematising global governance, democratisation of information ushering a post-truth era, technological advances risking the future of traditional manufacturing jobs, and global interconnectedness provoking socio-economic anxieties around the world.

Furthermore, the future of the liberal world order stood at a crossroad with the evolving US–China rivalry. The Chinese threat to American primacy gradually went beyond the specific realms of illiberal Chinese trade practices, Beijing's territorial aggrandisement in the South China Sea, and the expansion of China's regional influence at the hands of debt-trap diplomacy. Underpinned by its economic rise which oversaw the lifting of 850 million people out of poverty (since the 1980s),[1] the Chinese Communist Party's (CCP) techno-authoritarian model of governance seemed to have gained popularity – at the cost of a decline in the allure of the liberal democratic growth model. Certainly, the latter was also aided by the rise of populist strongmen leaders from Hungary to Brazil, governments from Egypt to Belarus engaging in overt securitisation to clamp down on dissent, and, finally, executive overextension via usurping of legislative authority by democratically elected governments from Turkey to the Philippines.

Accentuating the trend at the time, was also the US – the long-standing steward of liberal democracies, seemingly buckling under the pressures of rabid polarisation of the Barack Obama years. In many ways, the resultant hyper partisanship and rancorous public discourses only fed Donald Trump's rise. With his campaign mounting the takeover of the Republican Party, Trump successfully tapped into Americans' socio-economic anxieties at the hands of issues like comprehensive immigration reform and skewed trading arrangements awaiting resolution in a partisan and gridlocked Washington.

From a foreign policy standpoint, however, Trump's 'America First' prescription for US foreign policy – which expectedly would weaken US alliances from Western Europe to Northeast Asia, undermine US-led multilateral institutions, and cede the mantle of US advocacy of civil liberties – raised the possibility of an American retreat at a crucial time in its evolving rivalry with China and the future of the US-led liberal world order. So much so that, erstwhile bitter rivals – neoconservatives and liberal internationalists – on the ideological spectrum of US foreign policy found common ground in opposing Trump's proposition for American (dis)engagement from the world. In a series of open letters, former national security professionals, retired diplomats, and ex–cabinet chief from past Democratic and Republican administrations predicted Trump would be "the most reckless president in American history",[2] deemed him to be "ignorant of the complex nature of the challenges facing our country",[3] and even underscored the America First vision to be "wildly inconsistent and unmoored in principle".[4]

Furthermore, such concerns over the impact of Trump's America First worldview emerged, particularly with respect to the future of the Obama administration's Pivot – subsequently renamed Rebalance, to the Asia-Pacific policy, which sought to assert the US as a "Pacific power"[5] and shore up American defence, diplomatic, and economic ties in the region. After much focus on the Persian Gulf and the Middle East at large, Obama had sought to shift priorities of US foreign and security policy – in belated recognition of the need to counter China's rising influence.

However, Trump's campaign trail comments over deriding US security commitments towards Japan and South Korea – alliances which form the cornerstone of the US' "hub & spokes" alliance system in the region[6] – fed concerns over a prospective Trump administration undermining America's strategic edge in the region. At one point, Trump even fanned the possibility of a nuclear arms race in the region, when he suggested that Tokyo and Seoul should acquire their own nuclear weapons against regional threats instead of the US extending its security umbrella over them.[7] In addition, Trump's reported praise for the Philippines' strongman leader, Rodrigo Duterte's expansive war on drugs – which by end-2016 had left at least 4,500 dead[8] – raised concerns over a prospective Trump administration discontinuing the US foreign policy precedent of championing liberal values and rule of law. The dilemma over this was further compounded when then-candidate Trump praised North Korean leader Kim Jong-un's ascendency to power via his brutal campaign of eliminating political opponents. ("You gotta give him credit. . . . It's incredible. He wiped out the uncle, he wiped out this one, that one. I mean this guy doesn't play games".)[9]

Moreover, with heightened campaign rhetoric against China, many were apprehensive over a prospective Trump administration exacerbating the already "daunting challenge" of managing America's ties with its rising near-peer competitor.[10] Most importantly, given Trump's apparent abhorrence for multilateral trading arrangements such as the Obama administration's signature Trans-Pacific Partnership (TPP) for furthering a US-centric economic regional order, concerns emerged over a prospective Trump administration dropping the ball on America capitalising on the much-touted 'Asian Century'. After all, the region back then was already emerging as the world's new economic centre of gravity, with it being home to over half of the world's population, accounting for over one-third of global GDP and two of the world's three biggest economics hailing from Asia.[11]

Upon taking office in January 2017, Trump was quick to put into action some of his 'America First' prescriptions with his withdrawal from the TPP three days after taking office, announcing steel and aluminium tariffs on regional powers like China, Japan, Australia, Taiwan by making no apparent distinction between friends and foes, and pursuing direct negotiations with North Korea whilst undercutting allied South Korea. However, there were also evident points of continuity – as with further cultivating strategic partnerships in the region, employing corrective measures to the predecessor administration's approach (like adopting a sustained confrontational posture against China), and coaxing partner nations to increase burden-sharing.

Therefore, the Trump administration's record in the region was mixed at best – riddled with ample contradictions as well – with abhorring multilateralism yet purporting a geoeconomic vision under the newly christened "Indo-Pacific" region; decrying the role of values in US foreign policy whilst increasing the focus on China's civil liberties record; and, berating allied nations for their dependence on US security commitments but pursuing institutionalisation of strategic convergences with nascent partners like India.

Hence, this book aims to distil Trump's record in the Indo-Pacific region and further the scholarship on the impact of conservative nationalism on American internationalism.

In doing so, first, we assess the successes and failures of the Obama administration's Pivot/Rebalance policy, politicisation of America's role in Asia during the bitterly contested 2016 campaign, and the initial concerns over policy continuity in the region following Trump's election victory. Thereafter, we discuss the Trump administration's policy towards China and its role in shaping the emergent international – and domestic – consensus on addressing China's rise. Subsequently, an overview of the Trump administration's region-wide strategy probes its record on traditional and nascent partner nations, adherence to multilateralism, and the outlining of America's proposition under the "economic pillar" of the Indo-Pacific strategy. Finally, an evaluation of Trump's record on South Asia encompasses a look into his administration's policy towards Afghanistan, the resultant ebbs and flows of the US–Pakistan relationship, and the compartmentalisation of the US–India dynamic from events pertaining to the Af-Pak theatre and heightened bilateral trade tensions. Finally, we conclude with key takeaways from the Trump administration's record in the Indo-Pacific region and its implications for US foreign policy at large.

Notes

1 Yuen Yuen Ang, "Commentary: China Lifted 850 Million People out of Poverty but Now Faces Bigger Challenges", *Channel News Asia*, 22 November 2019, last accessed 15 September 2020, www.channelnewsasia.com/news/commentary/china-poverty-escape-economics-growth-inequality-nobel-prize-rct-12103676

2 Caroll Morello, "Former GOP National Security Officials: Trump Would Be 'Most Reckless' American President in History", *The Washington Post*, 08 August 2016, last accessed 15 September 2020, www.washingtonpost.com/world/national-security/group-of-50-former-gop-national-security-officials-trump-would-be-most-reckless-president-in-american-history/2016/08/08/6715042c-5d9f-11e6-af8e-54aa2e849447_story.html

3 Karen DeYoung, "75 Retired Senior Diplomats Sign Letter Opposing Trump for President", *The Washington Post*, 22 September 2016, last accessed 15 September 2020, www.washingtonpost.com/world/national-security/75-retired-senior-diplomats-sign-letter-opposing-trump-for-president/2016/09/21/5c5dff10–8046–11e6-b002–307601806392_story.html

4 Thomas Gibbons-Neff, "Trump Is 'Fundamentally Dishonest,' Say GOP National Security Leaders in Open Letter", *The Washington Post*, 03 March 2016, last accessed 15 September 2020, www.washingtonpost.com/news/checkpoint/wp/2016/03/02/in-new-letter-republican-foriegn-policy-experts-declare-war-on-trump/

5 Barack Obama, "Remarks by President Obama to the Australian Parliament", *The Barack Obama Administration – The White House Archives*, 17 November 2011, last accessed 15 September 2020, https://obamawhitehouse.archives.gov/the-press-office/2011/11/17/remarks-president-obama-australian-parliament

6 David Shambaugh, "China Engages Asia: Reshaping the Regional Order", *International Security*, Volume 29, Issue 3 (Winter 2004/05), p. 95.

7 *Quoted in* Maggie Haberman and David Sanger, "Transcript: Donald Trump Expounds on His Foreign Policy Views", *The New York Times*, 26 March 2016, last accessed 15 September 2020, www.nytimes.com/2016/03/27/us/politics/donald-trump-transcript.html

8 Emily Rauhala, "Duterte: During Phone Call, Trump Praised My Drug War as the 'Right Way'", *The Washington Post*, 03 December 2016, last accessed 15 September 2020, www.washingtonpost.com/news/worldviews/wp/2016/12/03/duterte-during-phone-call-trump-praised-my-drug-war-as-right-way/

9 *Quoted in* Evan McCurry, "Trump on North Korean Leader Kim Jong-un: 'You Gotta Give Him Credit,'" *ABC News*, 11 January 2016, last accessed 15 September2020,https://abcnews.go.com/Politics/trump-north-korean-leader-kim-jong-gotta-give/story?id=36198345

10 Eleanor Albert, "What Future for the Asia Pivot Under Trump?" *Council on Foreign Relations*, 14 December 2016, last accessed 15 September 2020, www.cfr.org/expert-roundup/what-future-asia-pivot-under-trump

11 White House factsheet, "FACT SHEET: Advancing the Rebalance to Asia and the Pacific", *The Barack Obama Administration – The White House Archives*, 16 November 2015, last accessed 15 September 2020, https://obamawhitehouse.archives.gov/the-press-office/2015/11/16/fact-sheet-advancing-rebalance-asia-and-pacific

1

OBAMA'S ASIA AND TRUMP'S CAMPAIGN

On inheriting two expansive Global War on Terror (GWOT) military efforts from the George W Bush administration, Barack Obama had committed to leading the US "in a new direction" during the 2008 presidential election.[1] Upon assuming office, in his attempt "to end the war in Iraq and properly prosecute the war in Afghanistan",[2] Obama often invoked the "just war" theory to support the latter[3] and explained his opposition to the former as him being "opposed to dumb wars".[4] As a result, Obama announced the withdrawal of the remaining 40,000 US troops in Iraq by the end of 2011 (as per the Status of Forces agreement signed under Bush).[5] Whereas, after conducting an administration-wide review of the US' approach in Afghanistan – which culminated in Obama announcing a surge of an additional 30,000 troops to Afghanistan – his administration set a predetermined withdrawal timeline of 18 months.[6]

With the US on path to extricating itself from the two wars which had come to be seen as a sign of American hegemonic excesses of the Bush years, the Obama administration sought to shift US priorities and capitalise on the much touted 'Asian Century'. In 2010, Obama's then-Deputy National Security Adviser Ben Rhodes said, "We see core U.S. national interests that will be advanced by us playing a key role in helping to shape the future of the region and making clear that we're an Asian and a Pacific power".[7] In asserting America's commitment to Asia – over which Obama had criticised his predecessor for paying "too little attention"[8] – Obama instituted a focus on the region in his administration's maiden year itself. Beyond declaring himself to be "America's first Pacific President",[9] in 2009, Obama became the first American president to meet all ten leaders of ASEAN member nations as a group,[10] and Hillary Clinton became the first US secretary of state in a generation to make Asia the destination of her first overseas visit.[11]

Thereafter, in 2010, the first signs of the administration's focus on the region taking shape with specific US policy priorities emerged. Speaking at the ASEAN Regional Forum in Hanoi, Clinton announced that the peaceful resolution of competing claims in the South China Sea was in the US' "national interest".[12] The subregion, which has been known to be rich in natural oil and gas deposits, has been subjected to competing claims. Chiefly, with China claiming a majority of the territory – which overlaps with the continental shelves of the Philippines, Brunei, Malaysia, Indonesia, Taiwan, and Vietnam. In her remarks, Clinton announced that Washington would support "a collaborative diplomatic process by all claimants for resolving the various territorial disputes without coercion" and even denounced the "use or threat of force by any claimant".[13] Her comments were deemed by China's foreign minister as "an attack on China",[14] since there was little doubt about the addressee of Clinton's remarks. With China construing its claims as a matter of "core interest" of its sovereignty,[15] in the past, Beijing had resorted to using force to seize parts of the contested area – such as its 1974 occupation of the western Paracels and subsequently the Mischief Reef in 1995.[16]

Shortly thereafter, the Obama administration formalised its focus towards the region with the announcement of the Pivot to Asia strategy. Writing in *Foreign Policy* magazine, Clinton made a case against giving in to retrenchment impulses following the winding down of Global War on Terror (GWOT) wars, by noting the criticality of Asia in "America's future" and "an engaged America" being "vital to Asia's future".[17] In outlining "a multifaceted and persistent effort" to actualise America's "irreplaceable role in the Pacific", Clinton outlined an approach based "along six key lines of action: strengthening bilateral security alliances; deepening our working relationships with emerging powers, including with China; engaging with regional multilateral institutions; expanding trade and investment; forging a broad-based military presence; and advancing democracy and human rights".[18]

Gains under the Pivot

Although Obama often criticised his predecessor for having "squandered" an opportunity with the GWOT effort to rally partner nations, the Bush administration's record in Asia reflected a slightly different reality. For instance, with Japan, the Bush administration laid the initial precedent of security cooperation. In addressing the Japanese Diet, Bush underscored the two nations' "common interests, common responsibilities and common values" and welcomed Japan "for providing important logistical support" to "rebuild a liberated Afghanistan".[19] For the first time since the end of the Second World War, Japan deployed its forces abroad towards assisting America's GWOT efforts. With the passage of the Anti-Terrorism Special Measures Law[20] – to ensure these exceptions don't overtly test the limits to Japan's offshore deployments under its post-war Pacifist Constitution – Tokyo deployed its Maritime Self-Defense Force in the Indian Ocean to provide refuelling assistance to NATO forces operating in Afghanistan.[21]

Similarly, GWOT accorded the US an opportunity to expand the scope of security cooperation with South Korea with its "out of area" contributions[22] – with Seoul's 2004 deployment of "1,400 combat Marines and Special Forces commandos and 1,600 military engineers and medics" to the "465 Korean military medics and engineers" already present in Iraq since 2003.[23] In addition, with Oceanic partners such as Australia, the first invocation of the ANZUS Treaty of 1951 led to significant contributions by Canberra towards US efforts in Iraq and Afghanistan.

In following this impetus to security cooperation between military forces of the US and its Asian and Oceanic partner nations, the Obama administration gained crucial latitude to then argue for recalibrating strategic ties under the announced Pivot to Asia policy.

For instance, in endorsing Clinton's postulations as per her *Foreign Policy* magazine article, Obama addressed the Australian Parliament in late 2011 to announce that the he had "made a deliberate and strategic decision – as a Pacific nation, the United States will play a larger and long-term role in shaping this region and its future".[24] In offering his "salute" to Australia for being "the largest contributor of troops" (outside NATO) to Afghanistan, Obama sought Canberra's support in the US now turning its "attention to the vast potential of the Asia-Pacific region".[25] In defining the future of US–Australia security cooperation, during the same visit, Obama announced the US' plan to oversee troop rotations to eventually build up to 2,500 US Marines being stationed at Darwin, Australia.[26]

In addition, the Obama administration enhanced relations with Tokyo by sealing a renewed set of guidelines for US–Japan defence cooperation. The same built on exceptions such as the discussed Anti-Terrorism Special Measures Law, to increase the possibility of Japan "signing on to U.S. military priorities elsewhere".[27] Most notably, the 2015 Revision of the Guidelines for US–Japan Defense Cooperation laid out frameworks for the two countries to "address bilateral cooperation on cybersecurity, the use of space for defense purposes, and ballistic missile defense".[28] Moreover, with the Establishment of the Alliance Coordination Mechanism (ACM), Japan and the US instituted an integrated command structure for contingencies involving common threats. This was a belated addition and a warranted upgrade in view of such measures existing in other American alliances frameworks, as with the North Atlantic Treaty Organization. This was the first such revision to US–Japan security guidelines since 1997 and was hailed by then-Secretary of State John Kerry as "an historic transition".[29] In addition, despite Chinese alarmism over Japan "re-militarizing", the Shinzo Abe government also reinterpreted Japan's post-war pacifist constitution to permit Tokyo to exercise the right to "collective self-defense".[30]

The Obama administration subsequently also renamed its Pivot to Asia effort to Rebalancing to the Asia-Pacific in order to emphasise that the United States "had never left Asia" and assert its status as a Pacific power.[31]

This made sense mainly from the standpoint of the military component of the Obama administration's plan of posturing American naval forces from the then "roughly 50–50% split between the Pacific and the Atlantic to about a 60–40 split between those oceans" to eventually encompass the shift of six aircraft carriers and a majority of US cruisers, destroyers, combat ships, and submarines.[32] In addition, advocating for America's "rebalance" to the region also included "rebalancing within the Asia-Pacific region"[33] – either in terms of pushing for increased burden sharing with partner nations or reducing the concentration of US forces in Northeast Asia, which dates back to the immediate aftermath of the Second World War and the subsequent Korean War. Hence, in modernising the US' military footprint in Northeast Asia, the Obama administration encouraged "a more "flexible" approach to deployments in the region", wherein American deployments would be "smaller, more agile, expeditionary, self-sustaining, and self-contained".[34] Thus, whilst underscoring the historical relevance and continued deterrence value of US forward positions – chiefly at Camp Humphreys and Osan Air Base in South Korea (ROK) and the Yokosuka naval base and Yokota Air Base in Japan, the Obama administration sought to further the model of rotational deployments to other strategically important locations like Australia, Singapore, Guam, and Hawaii. Moreover, in reducing partner nations' operational dependence on the US, the Obama administration in 2014 agreed with Seoul on "a conditions-based, rather than timeline-based, approach to determining when South Korea would obtain operational control of alliance forces in the event of a war".[35] Under negotiations of this Operational Control (OPCON) transition plan, the two parties agreed to recognise ROK's "advances in economic and military strength since the Korean War".[36]

Moreover, the Obama administration also sought to further security cooperation between Japan and South Korea. Historical enmity over outstanding issues had long plagued Japan–ROK ties, until the 2015 resolution of the comfort women issue – which witnessed Shinzo Abe issuing a formal apology and providing JPY 1 billion for a Korean foundation dedicated to supporting the surviving victims of that untoward practice by Japanese soldiers during the Second World War.[37] Reportedly, the Obama administration played a crucial role in facilitating the resolution of this issue, which came to be seen as a major breakthrough in one of the biggest impediments not only to Japan–ROK ties but also towards prospective US–Japan–ROK trilateral cooperation. Following which, the Obama administration successfully sought Tokyo and Seoul to take incremental steps towards bilateral security cooperation with the 2016 conclusion of the General Security of Military Information Agreement (GSOMIA) between Tokyo and Seoul.[38] The Obama years also witnessed the first ever trilateral ballistic missile warning exercise among the three nations.[39]

Beyond such developments on increased burden-sharing in Northeast Asia, the Obama administration, in its bilateral engagement with South Korea, witnessed crucial developments such as Seoul and Washington finalising the terms for a five-year Special Measures Agreement (covering 2014–18). In bringing its contribution to around 50 percent (by 2016) of the total cost of stationing US troops, the agreement encompassed South Korean commitments to year-on-year increases to its payments.[40] In addition, the two sides also signed a free trade agreement i.e. the Korea–US (KORUS) free trade agreement, the US announced the deployment of the Terminal High Altitude Area Defense (THAAD) system in South Korea to deter against North Korean provocations, and in 2015, with prodding by the Obama administration, South Korean cabinet officials outlined Seoul's position against China's actions in the South China Sea. South Korea even hosted the 2012 iteration of President Obama's flagship non-proliferation forum i.e. the Nuclear Security Summit.[41]

On Southeast Asia, the Obama years saw significant overtures like, Obama attending every US–ASEAN *and* East Asia summits from 2011 (apart from 2013 owing to a US government shutdown).[42] Furthermore, in a series of firsts:[43] In 2009, Obama signed the Treaty of Amity and Cooperation in Southeast Asia, which is known as "the anchor diplomatic document of ASEAN"; In 2010, the US became the first non-ASEAN country to "establish a dedicated Mission to ASEAN in Jakarta" and was represented by Secretary Clinton at the fifth East Asia Summit for the first time. In 2011, the Obama administration appointed the US' first resident ambassador to ASEAN and became the first country "to establish a dedicated Military Advisor/Liaison Officer" at the US Mission to ASEAN. In 2014, Secretary of Defense Chuck Hagel hosted his ASEAN counterparts in the US for the first US–ASEAN Defense Forum in Hawaii. In 2015, the administration elevated the relationship to a strategic partnership at the US–ASEAN Summit in Kuala Lumpur. Lastly, in 2016, the Obama administration hosted the Sunnylands Summit, which was hailed as the first US–ASEAN "stand-alone summit conducted in the United States".

In addition, in 2014, the US and the Philippines signed the Enhanced Defense Cooperation Agreement (EDCA) towards enhancing their 1951 Mutual Defense Treaty. Under the EDCA, US forces gained permission to utilise facilities "owned and controlled by the Armed Forces of the Philippines" – all whilst ensuring that the Philippine base commander "retains unhampered access to those locations".[44] As a result, the US military could build/improve the infrastructure of such installations, with local forces permitted to jointly use them. Under this provision, the Obama administration pursued joint development projects for the Philippines' capacity-building and thus oversaw unprecedented increases to US military aid to Manila. According to a report, the US boosted military aid from about US$ 50 million annually to $127.1 million in 2015–16 i.e. a 154 percent increase in military assistance from 2014–15 figures – accounting for the "biggest sum since American forces returned to the Philippines in 2002".[45] Similarly in 2015, with Singapore, the Obama administration signed an EDCA to expand bilateral security cooperation into "new areas, like humanitarian assistance and disaster relief, cyber defense, and counterterrorism".[46] This led to greater rotational deployments to Singapore such as the 2015 inaugural deployment of the US Navy's P-8A Poseidon maritime patrol aircrafts. This followed the US deploying up to four littoral combat ships in Singapore on a rotational basis starting 2013.

Lastly, the Obama administration deepened US relations with nascent partners. With Hanoi for instance, it spurred a dramatic reset in bilateral relations which oversaw the lifting of the US' decades-long arms embargo;[47] signing of the US–Vietnam Civil Nuclear Cooperation Agreement on transferring nuclear technology to the Southeast Asian nation;[48] and two American warships (submarine tender USS *Frank Cable* and guided-missile destroyer USS *John S McCain*) stopping in Cam Ranh Bay for the first time since resumption of US–Vietnam diplomatic relations.[49] Finally, with India, the Obama administration built on its predecessor administration's strategic outreach to the South Asian nation and accorded a fillip to US–India defence ties. In 2016, the Obama White House recognised India as a "Major Defense Partner" to facilitate "defense trade and technology sharing with the country on a level that the United States reserves for its closest friends and allies" and initiated the US–India Defence Technology and Trade Initiative to graduate US–India defence trade to a relation of co-production and co-development.[50] On interoperability, the Obama years also oversaw the upgrade of US–India Malabar naval exercise to a trilateral arrangement with Japan's inclusion in 2015, and the 2016 signing of the Logistics Exchange Memorandum of Agreement (LEMOA) for mutual logistics support between American and Indian forces.

Failings of the Pivot

Although the Obama administration substantially built on Bush's legacy of expanding American strategic partnerships in the region, it stopped short of articulating a region-wide engagement agenda. Although progress was made – especially with respect to the Obama administration's engagement with Southeast Asia via forums like ASEAN and the EAS, the Obama administration did not bind its actions in a narrative that went beyond merely calling to defend the status quo.

At first, there were signs of such an American proposition coming to the fore, with the Obama administration's TPP initiative for instance – wherein the US in no uncertain terms deemed it as an opportunity that "allows America – and not countries like China – to write the rules of the road in the 21st century".[51] However, in face of populist protests in partner nations like New Zealand – and the US itself[52] – the TPP's region-wide aim of instituting common labour and environmental standards and intellectual property protections[53] stood eclipsed by criticisms of its investor-friendly regulations which would permit legal actions against regional governments. With the continued push to see it through without addressing some of those criticisms over the deal undercutting participant nations' sovereignty, the multilateral arrangement accrued stigma as merely catering to the US policy aim of getting ahead of the subsequent China-backed Regional Comprehensive Economic Partnership (RCEP) – and thereby merely being aimed at solidifying US primacy.

This supposed revelation of the Obama administration's "true motive" in rebalancing to the Asia-Pacific – as China's Vice Foreign Minister termed it in 2012[54] – was also reflected in some other errors of the Obama administration. For example, in building on the Bush-era Pacific Vision war game to identify emergent security threats and focus on certain military skills that had "atrophied while the brunt of U.S. technical and tactical innovation was focused on Iraq and Afghanistan",[55] the Obama administration authorised the conceptualisation of the AirSea Battle (ASB) to overcome the challenges posed by Anti-Access, Area-Denial (A2/AD) threats. Despite it being a limited operational concept – warranting long-term investments in specific military capabilities (such as guided missiles and more silent submarines) and infrastructure in the cyber domain, for instance – the Obama administration openly touted it as being developed to counter China's military capabilities.

In warning against the open declaration of ASB being aimed at China, former Vice Chairman of the Joint Chiefs of Staff General James Cartwright in 2012 noted the ASB's effect of "demonizing China" to not be in "anybody's interest".[56] Subsequently, Section 1232 of the FY-2012 National Defense Authorization Act (NDAA) noted its purpose to also be towards combatting Iran's Anti-Access capabilities in the Persian Gulf. Further, in 2015, the Obama administration even renamed the same to the relatively benign – Joint Concept for Access and Maneuver in the Global Commons (JAM-GC), to illustrate a region-wide agenda behind the US' aims.[57] However, in 2017, John Ford – a captain in the US Army's JAG Corps, noted in hindsight that the ASB had already sent a signal of the "the entire pivot" being "part of a broader effort to encircle China".[58]

Furthermore, in a sign of the Obama administration being bogged down with the post–Cold War US policy dissonance of engaging and containing China, it also sent overtly conciliatory signals – which only raised questions over the US' commitment to the region and perplexed partner nations. In its attempt to reorient the US–China relationship by maximising cooperation with China on shared global governance challenges such as combatting climate change, the Obama administration sought to manage differences.

Most notably, in his visit to China in 2009, Obama was seen as "acquiescing" to China's positions in the region and ceding on American foreign policy precedents.[59] In Obama's joint statement[60] with Chinese President Hu Jintao, for instance, the two leaders affirmed their commitment that the "fundamental principle of respect for each other's sovereignty and territorial integrity is at the core of the three U.S.–China joint communiqués which guide U.S.–China relations". And the two sides agreed that "respecting each other's core interests is extremely important to ensure steady progress in U.S.–China relations". This was seen as Obama altering the American position on the 'One China' policy – according to which the US tactfully stopped short of accepting China's claim over Taiwan. In deeming "sovereignty and territorial integrity" to be at the core of US–China relations, the American position on continuing to build relations with China – whilst simultaneously continuing its security commitments to Taipei under the Taiwan Relations Act – seemingly stood altered to accepting the Chinese line of argument.[61] In addition, the joint statement also included a reference to both sides agreeing to "support the improvement and growth of relations between India and Pakistan". This was seen as China skilfully reinstating the hyphenation of India and Pakistan, contrary to the Clinton and Bush years' gradual – yet substantial – de-hyphenation of India and Pakistan in view of pursing a long-term American strategic partnership with New Delhi.

Similarly, in the years to come, even after announcing the US' pivot to the region, American allies and partner nations expressed doubts over the Obama administration's commitment to not undercut their interests. Such instances included the US supposedly looking to share power with China, as in case of the Obama administration's announced support in 2013 for China's "New Model of Great Power Relations" – a proposed "condominium of the United States and China in Asia" – feeding speculations of a coming G-2 power-sharing mechanism.[62] In addition, the Obama administration at times seemed rather satisfied in merely exacting verbal Chinese commitments, as in case of Chinese President Xi Jingping's 2015 Rose Garden commitment to not militarise the contested islands of the South China Sea.[63]

Finally, another issue which riddled Obama's actualisation of the pivot was the return of US security policy's focus on the Middle East. With American troops withdrawn from Iraq and poised to be drawn down in Afghanistan, the Obama administration had sought to shift its focus to the Asia-Pacific region. The Obama administration had pledged that, "after a decade of counterterrorism and wars in the greater Middle East, the United States – and the Department of Defense – would shift its investments, commitments, and operations to the Asia-Pacific".[64] However, despite the Obama administration seeking to underscore its focus on the Asia-Pacific region with the US president's tenth and final trip to Asia in 2016,[65] the gradual return of the US' focus on the Middle East was by then already evident with the Arab Spring movements upending the stability of US allies like Egypt, the US-led 2011 military intervention in Libya, Syria's civil war devolving into a transnational terrorism challenge with the rise of the Islamic State in Iraq and Syria (ISIS) in 2014, the US-backed Saudi Arabian intervention in the Yemeni civil war starting in 2015, and resumed hostilities between the navies of Iran and the US in 2016. On the diplomatic front, this shift in focus was also evident with Secretary Kerry's focus on diplomatically resolving the issue over Syria's stock of chemical weapons, cobbling a multination coalition (Operation Inherent Resolve) to fight ISIS, and bringing about the prolonged multiparty negotiations which ultimately led to the Joint Comprehensive Plan of Action (or the Iran Nuclear Deal).

Moreover, even before events in the Middle East warranted a change in US focus, Kerry was reported to have been sceptical of the military dimension of the pivot, which had already caused many to wonder "whether the rebalance to the Asia-Pacific was still an administration priority".[66]

For instance, in his confirmation hearing itself, Kerry downplayed the need to rebalance military presence to the region. He said, "I'm not convinced that increased military ramp-up is critical yet. . . . We have a lot more forces out there than any other nation in the world, including China today. . . . You know, the Chinese take a look at that and say, what's the United States doing? They trying to circle us? What's going on? And so, you know, every action has its reaction. . . . I think we have to be thoughtful about, you know, sort of how we go forward".[67]

Consequently, the second term of the Obama administration also witnessed the reduction of its "six key lines" approach – as postulated by Clinton,[68] to five "pillars". With Clinton's idea of "forging a broad-based military presence" evidently missing, National Security Advisor Thomas Donilon in 2013 outlined that approach to encompass: "strengthening alliances; deepening partnerships with emerging powers; building a stable, productive, and constructive relationship with China; empowering regional institutions; and helping to build a regional economic architecture that can sustain shared prosperity".[69] In addition, the "six key lines" approach was altered with the inclusion of the administration's priority to have "a stable and constructive relationship with China". Thus, even beyond Kerry casting the rebalance in a "action–reaction" narrative,[70] the administration downplaying the criticality of rebalancing military assets towards the region was further indicative of the continued relevance of the US' policy dissonance over engaging and containing China.

Continuity under question

In the bitterly contested 2016 US presidential campaign, concerns over the future of the Obama administration's Pivot – or Rebalance – to the Asia-Pacific emerged. With Donald Trump's focus on recalibrating America's trading relationships, commentators and policymakers alike worried about the impact of Trump's 'America First' worldview. In rallying against post–Cold War US bipartisanship on free trade, the Trump campaign vilified the American political class's culpability in spurring the hollowing-out of America's industrial base. Although Trump in large parts galvanised the white working class vote also by tapping into their socio-cultural anxieties over immigration – another policy agenda the campaign alleged the political establishment to have been ambivalent about – opposition to free trade was the cornerstone of the Trump campaign. In fact, in linking the two issues of trade and immigration, Trump accused the Washington establishment – or the "swamp" as he often called it on the campaign trail – of serving corporate interests.

For instance, in an extensive interview in 2019,[71] Steve Bannon – the primary ideologue behind Trump's 2016 conservative nationalist movement and subsequently chief strategist in the Trump White House, deemed the post–Cold War political establishment in Washington to have betrayed the American working class. In construing immigration and trade as "two sides of the same coin" and making no distinction between illegal immigration and uncontrolled legal immigration, Bannon accused the US elites to have come to terms with the US' "managed decline for unacceptable outcomes to average citizens". Bannon explained: "Mass illegal immigration is to flood the zone against predominantly black and Hispanic working class so that you've got . . . unlimited labor pool, and you can keep wages down for higher margins. Immigration and the H-1B visas are the same thing in the tech area, so that you don't have to hire American citizens". Ahead of the elections, analysts attributed the Trump campaign's focus on immigration as affecting US ties with Asian powers like India and the Philippines – that have particularly benefited from the corporate practice of outsourcing.

Further, Bannon explained the post–Cold War advocacy of free trade to be "the same thing" with the US "competing against foreign labor and foreign countries unfairly . . . to suppress workers' wages and to have higher margins; therefore, higher stock prices; therefore, more wealth, of which the workers don't own any piece of". The relevance of this thinking was apparent right from the start with Trump's announcement speech in 2015. In promising to "renegotiate our foreign trade deals", Trump said, "free trade can be wonderful if you have smart people, but we have people that are stupid. We have people that aren't smart. And we have people that are controlled by special interests. And it's just not going to work".[72]

This provided the Trump campaign with a substantial point of departure from the electoral proposition of Hillary Clinton – then-Democratic candidate for president. During her time in the US Senate [2001–09], Clinton had voted in favour of US free trade agreements with Singapore, Chile, Australia, Morocco, Bahrain, and Oman.[73] As Secretary of State, Clinton often underscored her and the Obama administration's commitment to free trade and played a crucial role in the United States–Colombia Trade Promotion Agreement (TPA) and the Korea–US (KORUS) free trade agreement – two trading arrangements which she had opposed during her 2008 presidential run. Moreover, as US secretary of state, she helped negotiate the Trans-Pacific Partnership (TPP) between 12 Pacific Rim nations and identified herself as an outspoken supporter of the same – and even called it the "gold standard" for future trading agreements.[74]

In a speech on his plan to rewrite global trade rules,[75] Trump even invoked the legacy of Clinton's husband and former US President Bill Clinton [1993–2001]. He said:

> America has lost nearly 1/3 of its manufacturing jobs since 1997. Even as the country has increased its population, think of this, by 50 million people. At the center of this catastrophe are two trade deals pushed by Bill and Hillary Clinton. First, the North American Free Trade Agreement, or the disaster called NAFTA. Second, China's entry into the World Trade Organization. NAFTA was the worst trade deal . . . . And China's entrance into the World Trade Organization has enabled the greatest job theft in the history of our country".

In addition, Trump construed Hillary Clinton to have "unleashed a trade war against the American worker when she supported one terrible deal after another, from NAFTA, to China to South Korea". On KORUS specifically, Trump alleged the deal to have "doubled our [US] trade deficit with South Korea and destroyed nearly 100,000 American jobs". Moreover, in finding common ground with Hillary's primary opponent and progressive candidate Bernie Sanders, Trump even quoted the independent senator from Vermont to invoke Clinton's voting record of supporting "virtually every trade agreement that has cost the workers of this country millions, millions of jobs".

Subsequently, as with her flip-flop on KORUS and TPA during the 2008 race, Clinton in the run-up to the 2016 election also criticised the TPP and mirrored Trump's prescription of imposing tariffs against Chinese goods.[76] In one of their early debates, Trump was quick to also pick up on this apparent political expedience to state: "And now you want to approve [the] Trans-Pacific Partnership. You were totally in favor of it! Then you heard what I was saying, how bad it is, and you said: 'I can't win that debate'. But you know that if you did win, you would approve that, and that will be almost as bad as NAFTA".[77]

Such instances permitted the Trump campaign to whip up conservative nationalist sentiment against the policies of "a leadership class that worships globalism".[78]

Moreover, given Clinton's centrality in the initiation of the Pivot to Asia campaign, the Obama administration's focus on the Asia-Pacific region also became politicised in the 2016 elections. Particularly on China, Trump deemed the "Obama-Clinton" foreign policy to have:

> allowed China to continue its economic assault on American jobs and wealth, refusing to enforce trade deals and apply leverage on China necessary to rein in North Korea. . . . He [Obama] has even allowed China to steal government secrets with cyber attacks and

> engaged in industrial espionage against the United States and its companies. We've let our rivals and challengers think they can get away with anything, and they do. They do at will. It always happens. If President Obama's goal had been to weaken America, he could not have done a better job.[79]

In then announcing his own vision for US trading arrangements, candidate Trump trained his sights not only on China, but also on US partner nations in Asia. In making no distinction between the United States' partner nations and near-peer competitors (like China) with his focus on renegotiating America's trading arrangements, Trump vowed: "We will not be ripped off anymore, we're going to be friendly with everybody, but we're not going to be taken advantage of by anybody".[80] On the US "losing vast amounts of billions of dollars" on underwriting the security of allied and partner nations, Trump even questioned the wisdom of the US' security partnerships. On the campaign trail, Clinton often warned against Trump's derision of American partner nations by assuring that her worldview valued "America's network of allies" as "what makes us exceptional".[81] Trump assumed the contrasting position of wondering what purpose such partnerships served. Specifically, on extending America's security umbrella over Japan and South Korea against North Korean brinksmanship, Trump asked: "Well, what are we getting out of this?"[82]

Subsequently, despite this heightened rhetoric on the Obama administration's failings and Clinton's supposed culpability in furthering American internationalism, there were signs that Trump was willing to take the baton on the US' policy aims in the region. For instance, even before he took office in January 2017, President-elect Trump met Japanese Prime Minister Shinzo Abe – the first foreign leader to have had a meeting with the incoming US president – and reaffirmed US security commitments towards Japan.[83] This instance was seen to have momentarily quelled fears over US policy continuity towards partner nations in the Asia-Pacific. On the other hand, there was evidence to suggest that Trump was willing to correct the Obama administration's overtly conciliatory approach towards China – albeit at the risk of upending long-standing precedents of US foreign policy. For instance, in an apparent sign of things to come under his campaign promise of "getting tough" with China, President-elect Trump spoke over the phone with Taiwanese President Tsai Ing-wen. This was the first time that an American president or president-elect had "spoken to a Taiwan president since 1979",[84] which is frowned upon under the unofficial dictums of the 'One China' policy. With the controversy that followed, President-elect Trump even said: "I fully understand the 'One China' policy, but I don't know why we have to be bound by a 'One China' policy unless we make a deal with China having to do with other things, including trade".[85]

Upon assuming office, such contradictions continued, as with Trump's withdrawal from the TPP three days after assuming office. Whereas, in many ways, Trump also carried forward the aims of the Pivot/Rebalance to

Asia-Pacific policy, when it devised a region-wide approach for the United States under the Indo-Pacific calculus. Before delving into this region-wide agenda which commendably encompassed furthering US partnerships much like how Obama built on Bush's legacy – the Trump administration's pursuit of a sustained confrontational agenda with China warrants discussion.

Notes

1 Barack Obama, "Obama's Remarks on Iraq and Afghanistan", 15 July 2008, last accessed 16 September 2020, www.nytimes.com/2008/07/15/us/politics/15text-obama.html

2 *Quoted in* Edward Delman, "Obama Promised to End America's Wars – Has He?", *The Atlantic*, 30 March 2016, last accessed 16 September 2020, www.theatlantic.com/international/archive/2016/03/obama-doctrine-wars-numbers/474531/

3 Barack Obama, "Full Text of Obama's Nobel Peace Prize Speech", *NBC News*, 12 October 2009, last accessed 16 September 2020, www.nbcnews.com/id/34360743/ns/politics-white_house/t/full-text-obamas-nobel-peace-prize-speech/#.Xyvi3i0Q1E4

4 *Quoted in* Helene Cooper, Shan Carter, et al., "On the Issues: Iraq and Afghanistan", *The New York Times*, n.d., last accessed 16 September 2020, www.nytimes.com/elections/2008/president/issues/iraq.html

5 BBC report, "US President Barack Obama Announces Iraq Withdrawal", *BBC*, 21 October 2011, last accessed 16 September 2020, www.bbc.com/news/av/world-us-canada-15410534

6 Barack Obama, "The New Way Forward – The President's Address", *The Barack Obama Administration – The White House Archives*, 01 December 2009, last accessed 16 September 2020, https://obamawhitehouse.archives.gov/blog/2009/12/01/new-way-forward-presidents-address

7 *Quoted in* Tom Cohen, "After Election Bruising, Obama Sets off on Extensive Asian Trip", *CNN*, 05 November 2010, last accessed 16 September 2020, https://edition.cnn.com/2010/POLITICS/11/05/obama.asia.trip/index.html

8 Kenneth Lieberthal, "The American Pivot to Asia", *Brookings Institution*, 21 December 2011, last accessed 16 September 2020, www.brookings.edu/articles/the-american-pivot-to-asia/

9 Barack Obama, "Remarks by President Barack Obama at Suntory Hall", *The Barack Obama Administration – The White House Archives*, 14 November 2009, last accessed 16 September 2020, https://obamawhitehouse.archives.gov/the-press-office/remarks-president-barack-obama-suntory-hall

10 Ernest Bower, "U.S.–ASEAN Summit: President Obama Engages Southeast Asia", *Center for Strategic and International Studies*, 09 November 2009, last accessed 16 September 2020, www.csis.org/analysis/us-asean-summit-president-obama-engages-southeast-asia

11 Jill Dougherty, "Clinton Heads to Asia on First State Trip", *CNN*, n.d., last accessed on September 16, 2020, https://edition.cnn.com/2009/POLITICS/02/15/us.clinton.asia.trip/

12 Hillary Clinton, "Remarks at Press Availability", *US Department of State – Archives*, 23 July 2010, last accessed 16 September 2020, https://2009-2017.state.gov/secretary/20092013clinton/rm/2010/07/145095.htm

13 Hillary Clinton, "Remarks at Press Availability", *The US Department of State – Archives*, 23 July 2010, last accessed 16 September 2020, https://2009-2017.state.gov/secretary/20092013clinton/rm/2010/07/145095.htm

14 *Quoted in* Gordon Chang, "Hillary Clinton Changes America's China Policy", *Forbes*, 28 July 2010, last accessed 16 September 2020, www.forbes.com/2010/07/28/china-beijing-asia-hillary-clinton-opinions-columnists-gordon-g-chang.html#2fb5ea1750c7
15 Mark Lander, "Offering to Aid Talks, U.S. Challenges China on Disputed Islands", *New York Times*, 23 July 2010, last accessed 16 September 2020, www.nytimes.com/2010/07/24/world/asia/24diplo.html
16 Gordon Chang, "Hillary Clinton Changes America's China Policy", *Forbes*, 28 July 2010, last accessed 16 September 2020, www.forbes.com/2010/07/28/china-beijing-asia-hillary-clinton-opinions-columnists-gordon-g-chang.html#2fb5ea1750c7
17 Hillary Clinton, "America's Pacific Century", *Foreign Policy*, 11 October 2011, last accessed 16 September 2020, https://foreignpolicy.com/2011/10/11/americas-pacific-century/
18 Hillary Clinton, "America's Pacific Century", *Foreign Policy*, 11 October 2011, last accessed 16 September 2020, https://foreignpolicy.com/2011/10/11/americas-pacific-century/
19 George W Bush, "President Discusses Unity Between the U.S. and Japan – Remarks by the President to the Diet", *The US Department of State – Archives*, 18 February 2002, last accessed 16 September 2020, https://2001-2009.state.gov/p/eap/rls/rm/2002/8616.htm
20 MOFA report, "Statement by Prime Minister Junichiro Koizumi on the Passing of the Anti-Terrorism Special Measures Law by the Diet of Japan", *Ministry of Foreign Affairs of Japan*, 29 October 2001, last accessed 16 September 2020, www.mofa.go.jp/region/n-america/us/terro0109/speech/pm1029.html
21 Stars and Stripes report, "Japan Ends Refueling Mission That Aided War Effort in Afghanistan", *Stars and Stripes*, 16 January 2010, last accessed 16 September 2020, www.stripes.com/news/japan-ends-refueling-mission-that-aided-war-effort-in-afghanistan-1.98083
22 CFR report, "Excerpt: The US–South Korea Alliance", *Council on Foreign Relations*, 2012, last accessed 16 September 2020, www.cfr.org/excerpt-us-south-korea-alliance
23 Balbina Hwang, "South Korean Troops to Iraq: A Boost for U.S.–ROK Relations", *The Heritage Foundation*, 13 February 2004, last accessed 16 September 2020, www.heritage.org/asia/report/south-korean-troops-iraq-boost-us-rok-relations
24 Barack Obama, "Remarks by President Obama to the Australian Parliament", *The Barack Obama Administration – The White House Archives*, 17 November 2011, last accessed 15 September 2020, https://obamawhitehouse.archives.gov/the-press-office/2011/11/17/remarks-president-obama-australian-parliament
25 Barack Obama, "Remarks by President Obama to the Australian Parliament", *The Barack Obama Administration – The White House Archives*, 17 November 2011, last accessed 16 September 2020, https://obamawhitehouse.archives.gov/the-press-office/2011/11/17/remarks-president-obama-australian-parliament
26 Jackie Calmes, "A U.S. Marine Base for Australia Irritates China", *The New York Times*, 16 November 2011, last accessed 16 September 2020, www.nytimes.com/2011/11/17/world/asia/obama-and-gillard-expand-us-australia-military-ties.html
27 CRS report, "The U.S.–Japan Alliance", *Congressional Research Service*, 13 June 2019, p. 7, last accessed 16 September 2020, https://fas.org/sgp/crs/row/RL33740.pdf
28 CRS report, "The U.S.–Japan Alliance", *Congressional Research Service*, 13 June 2019, p. 27, last accessed 16 September 2020, https://fas.org/sgp/crs/row/RL33740.pdf

29 *Quoted in* Arshad Mohammed, "U.S., Japan Unveil New Defense Guidelines for Global Japanese Role", *Reuters*, 27 April 2015, last accessed 16 September 2020, www.reuters.com/article/us-usa-japan-defense/u-s-japan-unveil-new-defense-guidelines-for-global-japanese-role-idUSKBN0NI08O20150427
30 Scott Harold, "President Obama's Visit to Japan and the Transformation of Japan's Defense Policy", *RAND Corporation*, 31 May 2016, last accessed 16 September 2020, www.rand.org/blog/2016/05/president-obamas-visit-to-japan-and-the-transformation.html
31 Richard Weitz, "Pivot Out, Rebalance In", *The Diplomat*, 03 May 2012, last accessed 16 September 2020, https://thediplomat.com/2012/05/pivot-out-rebalance-in/
32 BBC report, "Leon Panetta: US to Deploy 60% of Navy Fleet to Pacific", *BBC*, 02 June 2012, last accessed 16 September 2020, www.bbc.com/news/world-us-canada-18305750
33 Richard Weitz, "Pivot Out, Rebalance In", *The Diplomat*, 03 May 2012, last accessed 16 September 2020, https://thediplomat.com/2012/05/pivot-out-rebalance-in/
34 Mark Manyin, Stephen Dagget, et al., "Pivot to the Pacific? The Obama Administration's 'Rebalancing' Toward Asia", *Congressional Research Service*, 28 March 2012, last accessed 16 September 2020, https://fas.org/sgp/crs/natsec/R42448.pdf
35 Ash Carter, "The Rebalance and Asia-Pacific Security – Building a Principled Security Network", *Foreign Affairs*, November/December 2016, last accessed 16 September 2020, www.foreignaffairs.com/articles/united-states/2016-10-17/rebalance-and-asia-pacific-security
36 Mark Manyin, Emma Chanlett-Avery, et al., "U.S.–South Korea Relations", *Congressional Research Service*, 23 May 2017, p. 24, last accessed 16 September 2020, https://fas.org/sgp/crs/row/R41481.pdf
37 Mark Manyin, Emma Chanlett-Avery, et al., "U.S.–South Korea Relations", *Congressional Research Service*, 23 May 2017, p. 29, last accessed 16 September 2020, https://fas.org/sgp/crs/row/R41481.pdf
38 Reuters report, "South Korea, Japan Agree Intelligence-Sharing on North Korea Threat", *Reuters*, 23 November 2016, last accessed 16 September 2020, www.reuters.com/article/us-southkorea-japan-military-idUSKBN13I068
39 Wyatt Olson, "US, Japan, S. Korea Conducting First Joint Ballistic Missile Defense Drill", *Stars and Stripes*, 27 June 2016, last accessed 16 September 2020, www.stripes.com/news/us-japan-s-korea-conducting-first-joint-ballistic-missile-defense-drill-1.416554
40 Mark Manyin, Emma Chanlett-Avery, et al., "U.S.–South Korea Relations", *Congressional Research Service*, 23 May 2017, p. 23, last accessed 16 September 2020, https://fas.org/sgp/crs/row/R41481.pdf
41 Mark Manyin, Emma Chanlett-Avery, et al., "U.S.–South Korea Relations", *Congressional Research Service*, 23 May 2017, pp. 20–42, last accessed 16 September 2020, https://fas.org/sgp/crs/row/R41481.pdf
42 David Brunnstrom, "U.S. Set to Disappoint Asia with Downgraded Delegation for Bangkok Summits", *Reuters*, 30 October 2019, last accessed 16 September 2020, www.reuters.com/article/us-usa-asia/us-set-to-disappoint-asia-with-downgraded-delegation-for-bangkok-summits-idUSKBN1X9023
43 US in ASEAN report, "U.S.–ASEAN Timeline", *U.S. Mission to ASEAN*, n.d., last accessed September 16, 2020, https://asean.usmission.gov/our-relationship/policy-history/u-s-asean-timeline/
44 Renato Cruz de Castro, "Philippine Supreme Court Approves EDCA: Unlocking the Door for the Return of U.S. Strategic Footprint in Southeast Asia", *Center for Strategic & International Studies*, February 1, 2016, last accessed September 16, 2020, https://amti.csis.org/philippine-supreme-court-approves-edca-unlocking-door-return-u-s-strategic-footprint-southeast-asia/

45 Manuel Mogato, "As Alliance Wavers, U.S. Says Gave Philippines Big Annual Defense Aid Boost", *Reuters*, December 23, 2016, last accessed September 16, 2020, www.reuters.com/article/us-philippines-usa/as-alliance-wavers-u-s-says-gave-philippines-big-annual-defense-aid-boost-idUSKBN14C0QA
46 Barack Obama, "Remarks by President Obama and Prime Minister Lee of Singapore in Joint Press Conference", *The Barack Obama Administration – The White House Archives*, August 2, 2016, last accessed September 16, 2020, https://obamawhitehouse.archives.gov/the-press-office/2016/08/02/remarks-president-obama-and-prime-minister-lee-singapore-joint-press
47 BBC report, "Obama Lifts US Embargo on Lethal Arms Sales to Vietnam", *BBC*, May 23, 2016, last accessed September 16, 2020, www.bbc.com/news/world-asia-36356695
48 Lesley Wroughton, "US, Vietnam Sign Nuclear Trade Agreement", *Reuters*, October 10, 2013, last accessed September 16, 2020, www.reuters.com/article/us-usa-vietnam-nuclear/us-vietnam-sign-nuclear-trade-agreement-idUSBRE99904720131010
49 Bloomberg wire report, "United States Warships Make First Visit to Vietnam Base in Decades", *South China Morning Post*, October 4, 2016, last accessed September 16, 2020, www.scmp.com/news/asia/southeast-asia/article/2024978/united-states-warships-make-first-visit-vietnam-base
50 Ash Carter, "The Rebalance and Asia-Pacific Security – Building a Principled Security Network", *Foreign Affairs*, November/December 2016, last accessed September 16, 2020, www.foreignaffairs.com/articles/united-states/2016-10-17/rebalance-and-asia-pacific-security
51 Barack Obama, "Statement by the President on the Signing of the Trans-Pacific Partnership", *The Barack Obama Administration – The White House Archives*, 03 February 2016, last accessed 16 September 2020, https://obamawhitehouse.archives.gov/the-press-office/2016/02/03/statement-president-signing-trans-pacific-partnership
52 Caitlin McGee, "Controversial TPP Pact Signed amid New Zealand Protests", *Al Jazeera*, 04 February 2016, last accessed 16 September 2020, www.aljazeera.com/news/2016/02/controversial-tpp-pact-signed-zealand-protests-160204031601547.html
53 White House, "FACT SHEET: The United States in the Trans-Pacific Partnership – Increasing American Exports, Supporting American Jobs", *The Barack Obama Administration – The White House Archives*, 12 November 2011, last accessed 16 September 2020, https://obamawhitehouse.archives.gov/the-press-office/2011/11/12/fact-sheet-united-states-trans-pacific-partnership
54 *Quoted in* Susan Lawrence, "U.S.–China Relations: An Overview of Policy Issues", *Congressional Research Service*, 01 August 2013, last accessed 16 September 2020, https://fas.org/sgp/crs/row/R41108.pdf
55 Sam LaGrone, "Pentagon Drops Air Sea Battle Name, Concept Lives On", *USNI News*, 20 January 2015, last accessed 16 September 2020, https://news.usni.org/2015/01/20/pentagon-drops-air-sea-battle-name-concept-lives
56 *Quoted in* Amitai Etzioni, "Who Authorized Preparations for War with China?", *Yale Journal of International Affairs*, Summer 2013, p. 42.
57 Sam LaGrone, "Pentagon Drops Air Sea Battle Name, Concept Lives On", *USNI News*, 20 January 2015, last accessed 16 September 2020, https://news.usni.org/2015/01/20/pentagon-drops-air-sea-battle-name-concept-lives
58 John Ford, "The Pivot to Asia Was Obama's Biggest Mistake", *The Diplomat*, 21 January 2017, last accessed 16 September 2020, https://thediplomat.com/2017/01/the-pivot-to-asia-was-obamas-biggest-mistake/
59 Daniel Blumenthal, "Obama's Asia Trip: A Series of Unfortunate Events", *Foreign Policy*, 18 November 2009, last accessed 16 September 2020, https://foreignpolicy.com/2009/11/18/obamas-asia-trip-a-series-of-unfortunate-events/

60 White House press release, "U.S.–China Joint Statement", *The Barack Obama Administration – The White House Archives*, 17 November 2009, last accessed 16 September 2020, https://obamawhitehouse.archives.gov/realitycheck/the-press-office/us-china-joint-statement
61 Daniel Blumenthal, "Obama's Asia Trip: A Series of Unfortunate Events", *Foreign Policy*, 18 November 2009, last accessed 16 September 2020, https://foreignpolicy.com/2009/11/18/obamas-asia-trip-a-series-of-unfortunate-events/
62 Richard Bush, "The United States and China: A G-2 in the Making?", *Brookings Institution*, 11 October 2011, last accessed 16 September 2020, www.brookings.edu/articles/the-united-states-and-china-a-g-2-in-the-making/
63 Xi Jingping, "Remarks by President Obama and President Xi of the People's Republic of China in Joint Press Conference", *The Barack Obama Administration – The White House Archives*, 25 September 2015, last accessed 16 September 2020, https://obamawhitehouse.archives.gov/the-press-office/2015/09/25/remarks-president-obama-and-president-xi-peoples-republic-china-joint
64 Ash Carter, "The Rebalance and Asia-Pacific Security – Building a Principled Security Network", *Foreign Affairs*, November/December 2016, last accessed 16 September 2020, www.foreignaffairs.com/articles/united-states/2016-10-17/rebalance-and-asia-pacific-security
65 David Nakamura, "An Incomplete Victory Lap for Obama on His Final Presidential Trip to Asia", *The Washington Post*, 08 September 2016, last accessed 16 September 2020, www.washingtonpost.com/politics/no-victory-lap-for-obama-on-his-final-presidential-trip-to-asia/2016/09/08/fb9a56ce-75d2–11e6-b786–19d0cb1ed06c_story.html
66 David Berteau, Michael Green, and Zack Cooper, "Assessing the Asia-Pacific Rebalance", *Center for Strategic & International Studies*, December 2014, last accessed 16 September 2020, https://csis-website-prod.s3.amazonaws.com/s3fs-public/legacy_files/files/publication/150105_Berteau_AssessingAsiaPacificRebal_Web.pdf
67 *Quoted in* Max Fisher, "China Is Happy with John Kerry Because It Thinks He'll Drop the 'Pivot to Asia'", *The Washington Post*, 28 February 2013, last accessed 16 September 2020, www.washingtonpost.com/news/worldviews/wp/2013/02/27/china-is-happy-with-john-kerry-because-it-thinks-hell-drop-the-pivot-to-asia/
68 Hillary Clinton, "America's Pacific Century", *Foreign Policy*, 11 October 2011, last accessed 16 September 2020, https://foreignpolicy.com/2011/10/11/americas-pacific-century/
69 Thomas Donilon, "Complete Transcript: Thomas Donilon at Asia Society New York", *The Asia Society*, 11 March 2013, last accessed 16 September 2020, https://asiasociety.org/new-york/complete-transcript-thomas-donilon-asia-society-new-york
70 Elizabeth Economy, "Secretary of State John Kerry on China", *Council on Foreign Relations*, 27 February 2013, last accessed 16 September 2020, www.cfr.org/blog/secretary-state-john-kerry-china
71 *Quoted in* PBS transcript, "Steve Bannon – First Interview with the Former Trump Adviser", *Frontline – Public Broadcasting Service*, 17 March 2019, last accessed 16 September 2020, www.pbs.org/wgbh/frontline/interview/steve-bannon-3/
72 *Quoted in* TIME transcript, "Here's Donald Trump's Presidential Announcement Speech", *TIME*, 16 June 2015, last accessed 16 September 2020, https://time.com/3923128/donald-trump-announcement-speech/
73 Linda Qiu, "Clinton Voted for 'Virtually Every' Trade Agreement, Killing Millions of Jobs, Sanders Says", *Politifact*, 10 April 2016, last accessed 16 September 2020, www.politifact.com/factchecks/2016/apr/10/bernie-s/clinton-voted-virtually-every-trade-agreement-kill/

74 *Quoted in* Clark Packard, "Is Hillary Clinton a Free Trader?", *The Hill*, 26 July 2016, last accessed 16 September 2020, https://thehill.com/blogs/congress-blog/presidential-campaign/289163-is-hillary-clinton-a-free-trader
75 *Quoted in* TIME transcript, "Read Donald Trump's Speech on Trade", *TIME*, 28 June 2016, last accessed 16 September 2020, https://time.com/4386335/donald-trump-trade-speech-transcript/
76 Heather Long, "Clinton Suddenly Sounds a Lot Like Trump on Trade", *CNN*, 11 August 2016, last accessed 16 September 2020, https://money.cnn.com/2016/08/11/news/economy/hillary-clinton-trade/index.html
77 C-Span User Clip, "Trump v Clinton on Trade", *C-Span*, 27 September 2016, last accessed 16 September 2020, www.c-span.org/video/?c4622013/user-clip-trump-clinton-trade
78 *Quoted in* Nick Corasaniti, Alexander Burns, and Binyamin Appelbaum, "Donald Trump Vows to Rip Up Trade Deals and Confront China", *The New York Times*, 28 June 2016, last accessed 16 September 2020, www.nytimes.com/2016/06/29/us/politics/donald-trump-trade-speech.html
79 *Quoted in* NYT transcript, "Transcript: Donald Trump's Foreign Policy Speech", *The New York Times*, 27 April 2016, last accessed 16 September 2020, www.nytimes.com/2016/04/28/us/politics/transcript-trump-foreign-policy.html
80 *Quoted in* Maggie Haberman and David Sanger, "Highlights from Our Interview with Donald Trump on Foreign Policy", *The New York Times*, 26 March 2016, last accessed 16 September 2020, www.nytimes.com/2016/03/27/us/politics/donald-trump-interview-highlights.html
81 *Quoted in* Katie Reilly, "Read Hillary Clinton's Speech on Donald Trump and National Security", *TIME*, 02 June 2016, last accessed 16 September 2020, https://time.com/4355797/hillary-clinton-donald-trump-foreign-policy-speech-transcript/
82 *Quoted in* NYT report, "Transcript: Donald Trump on NATO, Turkey's Coup Attempt and the World", *The New York Times*, 21 July 2016, last accessed 16 September 2020, www.nytimes.com/2016/07/22/us/politics/donald-trump-foreign-policy-interview.html
83 MOFA press release, "Prime Minister Abe Held a Meeting with the President-Elect of the United States", *Ministry of Foreign Affairs of Japan*, 18 November 2016, last accessed 16 September 2020, www.mofa.go.jp/mofaj/na/na1/us/page3e_000622.html
84 Max Boot, "Donald Trump's Pivot Through Asia", *Foreign Policy*, 27 December 2016, last accessed 16 September 2020, https://foreignpolicy.com/2016/12/27/the-pivot-to-asia-obama-trump/
85 *Quoted in* Caren Bohan and David Brunnstrom, "Trump Says U.S. Not Necessarily Bound by 'One China' Policy", *Reuters*, 11 December 2016, last accessed 16 September 2020, www.reuters.com/article/us-usa-trump-china/trump-says-u-s-not-necessarily-bound-by-one-china-policy-idUSKBN1400TY

2

THE TRUMP PRESIDENCY AND CHINA

A sustained confrontational posture against China emerged as one of the key shifts in US foreign policy under the Trump administration. As in case of the administration's continuity on the need to shift US focus to the Indo-Pacific region, a degree of Obama's approach on China also figured in Trump's conduct of the US' most consequential bilateral relationship.

Under the Obama administration, the US levied a stiff 35 percent tariff on Chinese tires after US companies complained about unfair competition.[1] Much like Trump's 'tariff war' against China, the Obama administration in its maiden year acted against China's flooding of the international and American market with low-priced tires. This had made it difficult for American companies to compete, as their percentage share in the domestic and global market slipped, threatening nearly 1,200 manufacturing jobs in the industrial mid-West.[2] President Obama, in his 2012 State of the Union address, even hailed the results of his confrontational posture against China: "Over a thousand Americans are working today because we stopped a surge in Chinese tires".[3] However, this confrontational posture under Obama was short-lived.

Over time, with the emergence of the Tea Party movement, the Democrats' hold over the US Congress considerably weakened in the Obama years. Although it is a norm for parties that hold the White House to lose power in midterms, the Obama years led to the Democrats' precipitous slide. By the end of his terms, they had lost 11 Senate seats, 62 House seats, 12 governorships, and 958 seats in state legislatures.[4] This loss was greater than the Democrats' slide during the JFK/Johnson years' domestic instability and the Vietnam War, and even the Clinton years' surge of Newt Gingrich conservatives and the Monica Lewinsky scandal. Thus, after some major successes on the domestic front – like the passage of stimulus/bailout packages following the 2007–08 financial crisis and the signing of the Affordable Care Act into law, the Obama administration oversaw increased partisan gridlock on Capitol Hill.

In face of Democrats' declining hold over the two chambers of Congress, the Obama administration's domestic agenda faced increased obstructionism at the hands of the GOP's idea of, "We're going to do everything – and I mean everything we can do – to kill it, stop it, slow it down, whatever we can" – in Republican Speaker John Boehner's words.[5] Subsequently, even on pressing agendas like that of gun control amidst shootings spiking in schools and other public areas, Obama's pursuits on the Hill came to a screeching halt under the hyper-partisan GOP leadership of Speaker Paul Ryan.

This spurred Obama's overt reliance on executive authority on many domestic issues and a significant pivot to foreign policy matters – a realm under which the US executive branch enjoys considerable autonomy. With focus on big-ticket foreign policy issues like the finalisation of the Iran Nuclear Deal and the Paris Climate Accords, cooperation with China became a precursor given its seminal role in breaking the impasse over the Arak Heavy Water facility in case of the former[6] and its centrality as the largest source of global emissions in case of the latter.[7]

Hence, Obama's appetite for confronting China decreased, leading to the tire tariffs gradually waning and eventually ending in 2012.[8] This was followed by US capitulations on other fronts, such as ignoring China's continued expansionism in the South China Sea. This was in direct contradiction to Chinese President Xi Jingping's 2015 pledge against militarising those islands, in the presence of President Obama during a joint press conference in the White House Rose Garden.[9]

Hence, Trump's sustained confrontation against China was also aided by his abhorrence of multilateralism – which concurrently reduced the imperatives of US cooperation with China on pressing matters of global governance. However, the Trump administration's mandate on China ran deep – best evidenced in the ideas that bedrocked his 2016 presidential run.

China in American conservative nationalism

In Donald Trump's populist takeover of the Republican Party, his 2016 campaign linked Americans' socio-cultural anxieties on immigration and free trade to the US political elites' expedience on China. The Trump campaign galvanised white working-class voters by attributing the rising influx of immigrant workers to the US political establishment's wilful disinterest in pursuing comprehensive immigration reform. Furthermore, in attacking the US political establishment's post–Cold War bipartisan consensus on free trade, Trump alleged foreign countries to have "taken advantage" of the US.[10] During the campaign and thereafter, Trump purported the US to have lost 4.2 million manufacturing jobs and to have accrued US$ 15 trillion in trade deficits during the past quarter century.[11] In construing these ills of US support for free trade as having further compounded American workers' plight, Trump singled out the US' bilateral relationship with China – the Asian country where many US manufacturers had shifted their production base. Alleging past US policymakers to have been complicit in a scheme of orchestrating the US' "managed decline", in view of commercial interests on trade with China,[12] Trump would often point to the US' then–trade deficit of over US$ 350 billion with China. Thus, in making China a centrepiece of its narrative, the Trump campaign implied a causation: the US political establishment's malfeasance on trade with China had borne "unacceptable outcomes" for average citizens – in terms of job losses in the American Midwest, by encouraging "the shipping of its [US] industry base to China".[13]

Beyond effectively tapping into such nativist sentiment, the reason why this narrative resonated well with the American electorate was also the fact that the US political establishment had indeed bungled China's integration into the liberal world order.

In putting the proverbial cart before the horse, US policymakers in the late 1990s rallied around a bipartisan effort to tap into the huge potentialities of China's market, ahead of its liberal market reforms. For instance, as in case of the Obama administration, the Bill Clinton administration also began with a confrontational posture with China – albeit in that case human rights had been the major bone of contention owing to then-recent Chinese state crackdown on the 1989 Tiananmen Square protests. As a result, under the Jackson-Vanik Amendment to the 1974 Trade Act, the US political establishment necessitated the Clinton executive branch to annually review the US' decision to continue according China the Most Favoured Nation (MFN) status on account of its progress on civil liberties and human rights issues.[14] Eventually, in order to cultivate China as a "responsible stakeholder"[15] – primarily in recognition of China's massive market potentialities – the administration repealed the Jackson-Vanik Amendment's clause on China and passed the Permanent Normal Trade Relations (PNTR) legislation that ended the "annual ritual of reviewing China's trade status".[16]

Thereafter, the Republican administration of George W. Bush not only continued the PNTR policy, but also went a step further by facilitating China's entry into the World Trade Organisation (WTO).[17] Without the pre-PNTR policy of first necessitating socio-political reforms, this permitted China to continually kick the proverbial can down the road by brandishing multinational companies' access to their market (by then already the largest in the world) as a bargaining chip.

This bipartisan development in US policy towards China only weakened American resolve on necessitating Chinese liberal economic reforms. Although the Clinton and Bush administrations imposed some anti-dumping duties on Chinese imports, they failed to address the fundamental forces behind Chinese unfair trade practices.[18] Following this, in its gradual ascent as the world's second-largest economy and the largest trading partner of nearly all of the world's industrialised nations, China continued to peg its currency at artificially low levels, incentivise state-owned enterprises, and necessitate technological transfers in exchange for foreign manufactures entering its market.

Over time, as US–China bilateral trade grew, the debate over the ideal American policy towards China stood marred with confusion. Subsequent post–Cold War US administrations struggled over this policy dissonance – whether US commercial gains via cooperation with China outweighed the US' need to pursue containment over divergent strategic goals with China. This confounding question in US foreign policy was best reflected in a slip-up in one of President Bush's public addresses. In an apparent characterisation of his administration's dilemma over crafting an apt approach towards China, Bush seemed to be coming down on the side of short-term economic gains. He said, "China is a strategic partner, I mean, strategic competitor but that doesn't mean we can't find areas in which we can partner and the economy is a place we can partner".[19]

Drawing on such precedents in US foreign policy, Trump then committed to a conservative nationalist approach to rectify the impact of this cooperation–containment US policy dissonance. This, despite the fact that research suggests China's effect on US industrial manufacturing – in terms of losses in American jobs has been *relatively* lower than the impact that technology-driven, automated production has had.[20] Similarly, on China's rise leading to a relative decline in the US' economic primacy, benefits for the US economy have often been understated. For instance, after facilitating China's entry into the WTO, by the end of President Bush's two terms, American "merchandise exports to China increased 187 percent" and US exports "to the rest of the world grew 38 percent".[21]

However, as witnessed in populist movements around the world, the appeal of nationalism has been in its binary, reductionist characterisations of contemporary socio-economic trends that are often far too complex and grey to be distilled into short bumper-sticker campaign slogans. Thus, in Trump's case as well, the causation between the US' unfettered push for globalist ideas of free trade and legal immigration, and the post–Cold War US policy dissonance on China, worked for Trump amidst declining support for American internationalism and heightened socio-cultural anxieties. For instance, in 2016, a staggering 70 percent of the US electorate deemed immigration to be "very important" to their voting decision,[22] and 57 percent agreed with the statement that the US should "deal with its own problems and let others deal with theirs the best they can".[23]

As a result, with his victory, President Trump was then armed with the mandate to put to rest the long-standing US policy dissonance on China, and given the centrality of China in the 2016 fillip to conservative nationalism, an American policy of sustained confrontation emerged.

The 'America First' policy on China

The centrality of China in the 2016 presidential campaign translated into the US–China relationship becoming the centrepiece of the Trump administration's foreign policy. The administration's worldview and, especially, its view of China's rise were most prominently captured in its inaugural draft of the National Security Strategy (NSS).

Released in late 2017, it announced the re-emergence of "great power competition", with China and Russia wanting "to shape a world antithetical to U.S. values and interests". Additionally, it deemed them to be "fielding military capabilities designed to deny America access in times of crisis and to contest our ability to operate freely in critical commercial zones during peacetime. In short, they are contesting our geopolitical advantages and trying to change the international order in their favor".[24]

Beyond this open declaration of American primacy being under threat at the hands of China's rise, the Trump administration sought to institute a policy of sustained confrontation with its Asian near-peer competitor in the following three domains.

1. Weaponising the US' trade deficit to wage a tariff war

During the 2016 campaign and thereafter, Trump rallied against the US' nearly US$ 800 billion trade deficit with other nations.[25] Once in office, Trump even held deficits as a metric for judging America's trade relationships and as a result pushed for "fair and reciprocal" trade agreements.[26] By announcing tariffs on imports from countries – friends and foes alike – Trump weaponised America's trade deficits – the very thing he used to rally against the Washington elites' post–Cold War consensus on free trade. In his 2019 speech at the UN, he even deemed globalism to have "exerted a religious pull over past leaders, causing them to ignore their own national interests".[27]

With China especially, Trump argued for the merits of engaging in a trade war – which the WTO deemed to be an unarguable affront to global trade norms.[28] In announcing his March 2018 imposition of 25 percent tariff on steel and a 10 percent tariff on aluminium, Trump said: "When a country [USA] is losing many billions of dollars on trade with virtually every country it does business with, trade wars are good, and easy to win".[29]

In response, China imposed tariffs of up to 25 percent on 128 American products – including airplanes and soybeans. Thereafter, in the subsequent months, Trump's attempt to levy punitive tariffs on Chinese imports devolved into a game of chicken – i.e. over which side could hold on longer with their imposition of tariffs and retaliatory tariffs. All through 2018 and 2019 – albeit with some brief episodes of mutually agreed halts (as in December 2018 and August 2019) – both sides raised the stakes.[30] Many a times, US actions met proportional retaliations – as in August 2018, when the US released a list of US$ 16 billion of Chinese goods to be tariffed at 25 percent, and China retaliated with its own list of US$ 16 billion worth of US goods with 25 percent duties. In some instances, Trump raised the stakes at a considerable scale – as when trade talks stalled at the G20 meet in Osaka, Trump announced 10 percent tariffs on US$ 300 billion worth of Chinese imports. Cumulatively, the US imposed tariffs on more than US$ 360 billion of Chinese goods, and China retaliated with tariffs on more than US$ 110 billion of American products.[31]

This prolonged over a period of 18 months and at one point even began to reap counterintuitive results for the US – either in terms of American manufacturers (who import Chinese materials in order to then export finished goods back to China) bearing the brunt of tariffs on both ends or US farmers struggling at the hands of China curtailing the import of US agricultural goods (chiefly, soybean). On the latter, the US' total agricultural goods export to China, which was US$ 24 billion in 2014, fell to US$ 9.1 billion in 2019.[32] In order to then offset this damage to one of his most loyal electoral bases, Trump even announced farmer bailouts amounting to nearly US$ 30 billion.[33]

Finally, in late October 2019, the US and China announced a breakthrough in their trade negotiations. With little on structural reforms over its state subsidies or forced transfer of technology, the Phase One agreement encompassed Chinese assurances to raise imports from the US. Hence, on Trump's policy aim of narrowing America's trade imbalance, the January 2020 deal exacted a pledge from China to "increase imports of American goods and services by at least $200 billion" over the next two years – above a baseline of $186 billion in purchases in 2017.[34] On agricultural goods, that meant additional US exports to China of $12.5 billion in 2020 and $19.5 billion in 2021 (over the 2017 baseline of $24 billion).[35] Other commitments included US$ 54 billion in energy purchases, US$ 78 billion in manufacturing purchases, and US$ 38 billion in services.[36]

2. *Raising the significance of US FONOPS in China's backyard*

In accordance with greater operational latitude to US combatant commanders around the world, Trump also gave considerable leeway to the US Navy on its actions in the South China Sea. In July 2017, his administration deliberated "a full-year schedule of when US navy ships will sail through contested waters in the South China Sea".[37] Submitted by then-Defense Secretary James Mattis, the schedule sought to make the Trump administration aware of all planned "freedom of navigation operations" (FONOPs), so that requests for the same through the chain of command do not stand hindered by tardy processes. Moreover, from the standpoint of US–China competition on the high-seas, this meant a policy departure with the US looking to tout such operations as being conducted on a "very routine, very regular" basis rather than as a "one-off event" or as reactive measures against Chinese excesses.[38] A month prior to the announced plan, the US Navy had conducted its first freedom of navigation operation under the Trump administration. Wherein, the USS *Dewey* sailed within 12 nautical miles of Mischief Reef – part of the Spratly Island chain on which China has built airstrips – and carried out a 'man overboard' drill as it passed the region.[39]

In contrast to about half-a-dozen such operations under the Obama administration, the Trump administration by mid-2019 had already overseen that many visits of US Navy ships to the South China Sea as part of its FONOPs initiative.[40] By 2020, US Navy had conducted more FONOPSs in 2019 than "in any year since the U.S. began more aggressively challenging China's claims in the South China Sea in 2015".[41] While the Obama administration authorised two FONOPs in 2015 and three in 2016, the Trump administration stepped up the frequency of such operations with six in 2017, five in 2018, and nine in 2019.[42] According to Stephen Nagy (International Studies Professor, International Christian University, Tokyo), the Trump administration adopted "a clear consensus that the Obama administration didn't do enough of the FONOPs and that the increased number and breadth of the FONOPs is representative of a general shift in Washington".[43] Even amidst the COVID19 pandemic, as China pressed on with announcing administrative jurisdictions over the militarised islands, the US Navy conduced back-to-back missions in order to further the Pentagon strategy of "strategic predictability, operational unpredictability".[44] Similarly, the US also stepped up its transits through the Taiwan Straits: with 12 in 2016, five in 2017, three in 2018, and nine in 2019.[45]

Moreover, the heightened frequency of such US Navy operations in the Pacific seemed to have stemmed from an added rationale. With the Obama administration's limited authorisations of such operations, they came to be viewed as US responses to Chinese actions –aiming to possibly spur a change in Chinese behaviour in the South China Sea. That line of thought holds credence, especially if one considers the historical precedent of the US conducting transits through the Taiwan Straits during the Taiwan Straits Crises of the 1990s and the Chinese successfully backing down as a result.

However, since then, the asymmetry between American and Chinese capabilities has narrowed, especially at the hands of the latter's Anti-Access, Area-Denial (A2AD) capabilities. Hence, the efficacy of increased US FONOPs cannot merely be limited to the change it instigates in Chinese behaviour. Instead, as Bonnie Glaser (Director, China Power Project, CSIS) explains, the aim is to ensure "the US military can fly, sail, and operate wherever international law allows" and not just "narrowly assert the rights of foreign military vessels in accordance with the UN Convention on the Law of the Sea".[46] Simply put, the US sought to construe the same not "as noteworthy events, but more as a fact of life in the South China Sea – a reminder of the U.S. Navy's forward presence in the area and its commitment to freedom of navigation".[47]

3. Stemming the tide against Chinese technological primacy

Another facet of the Trump administration's sustained confrontation pertained to China's gradual rise as a technological power. In 2020, China became the biggest source of applications for international patents at the World Intellectual Property Organization (WIPO) – dislodging the US' streak of holding the top spot every year since the Patent Cooperation Treaty came into force in 1978.[48] With China's applications seeing a 200-fold increase in the past 20 years, according to WIPO, Huawei Technologies has been the top corporate patent filer for the past three consecutive years.[49] Owing to this and considerable support from the Chinese state, Huawei has emerged as the world's biggest manufacturer of telecom equipment. Moreover, with regard to the next frontier of telecommunications, i.e. 5G network and its concurrent impetus to the 'Internet of things', Huawei has touted itself as the only end-to-end equipment producer of 5G telecommunications network, over competitors such as Finland's Nokia and Sweden's Ericsson.

In response, by citing a 2017 Chinese intelligence law that mandates Chinese companies to cooperate in state intelligence work, the Trump administration initiated a global campaign warning countries against permitting Huawei to participate in their respective trials for 5G networks. At home, the administration even sought to decrease interlinkages between American entities and Huawei. For instance, in 2019, the Trump administration placed Huawei on a trade blacklist, banning it from doing business with US firms. That meant it could no longer "buy hardware components like semiconductors or access critical software such as Google services on Android, putting its smartphone business outside China at a huge disadvantage".[50]

Further, the Trump Justice Department announced indictments to charge the telecommunications giant for using "fraud and deception to misappropriate sophisticated technology from U.S. counterparts", starting a "policy instituting a bonus program to reward employees who obtained confidential information from competitors", and making "repeated misstatements to U.S. officials, including FBI agents and representatives from the U.S. House Permanent Select Committee on Intelligence, regarding their efforts to misappropriate trade secrets".[51]

In addition, the Federal Communications Commission voted unanimously to classify Huawei as a "national security threat" and barred American telecom companies from using FCC funds to purchase equipment from such entities.[52] Finally, Trump's Attorney General William Barr also suggested considering American ownership or a controlling stake of alternate options like Nokia and Ericsson: "Putting our large market and financial muscle behind one or both of these firms would make it a more formidable competitor and eliminate concerns over its staying power".[53]

With the US gradually decoupling its tech sector from Huawei and raising alarm amongst its partner nations, US public opinion moved as well.

For instance, according to a poll in mid-2020, nearly half of US adults (47 percent) said that they are concerned about China's growing technological power.[54]

However, with partner nations only gradually heeding to US arguments and effectively banning Huawei from their respective 5G trials, there were concerns if the US' focus on Huawei is a mere bargaining chip in the broader US–China confrontation over market access and China's trade practices. For instance, in the early days of his administration, Trump similarly sparred with another Chinese tech company, ZTE Corp. But after its three-month-long ban, Trump removed US sanctions after the Chinese heeded Trump's call for revisiting US–China trade terms. The company reached a settlement with Trump's Commerce Department by paying a US$ 1 billion penalty and a deposit of US$ 400 million as escrow and by agreeing to institute a compliance team answerable to the department.[55]

In Huawei's case, however, given its centrality in China consolidating its technological primacy, the Trump administration underscored its unwavering political will. For instance, even as China and the US finalised the Phase One agreement, the Trump administration did not shy from raising the pressure on Huawei. In May 2020, for instance, the Commerce Department announced tightened regulation to prevent Huawei and its subsidiaries from using American technology and software. Essentially, the rule was aimed to "block companies around the world from using American-made machinery and software to design or produce chips for Huawei or its entities".[56] In addition, in September 2020, the Trump administration even announced its intent to bar people in the US from downloading Chinese-owned apps WeChat and TikTok.[57]

COVID19 and American bipartisanship on China

The coronavirus pandemic expectedly served as a shot-in-the-arm for anti-globalisation forces around the world. Moreover, amidst rising nativist impulses against the free movement of people – with tighter tourism regulations possibly paving the way for xenophobic anti-immigration laws – liberal economic dependence also seemed to come into question.

With the pandemic already exacerbating socio-cultural nativism and economic nativism, it spurred a realisation amongst nations over their level of dependence on China. This pertained to dependencies ranging from utility goods like tires – China is the largest producer of tires with an estimated 759.3 million tire units manufactured in 2019[58] – to crucial raw materials like Active Pharmaceutical Ingredients (APIs) – with Chinese manufacturers accounting for 40 percent of the global requirement for those chemical formulations that are central to the production of common drugs.[59] Simply put, the coronavirus pandemic catalysed the world's rude awakening to China's centrality in modern transitional commerce, with Beijing accounting for roughly one-sixth of the global economic output as the world's largest manufacturer.[60]

In addition, China's centrality to other economies goes beyond its standing as a major supplier. As the most populous nation on the planet, its scale of consumption is key to many transnational businesses' ledger books. For instance, apart from being the largest automobile and smartphone market, Chinese tourists outspend US tourists by a factor of two with US$ 258 billion per year.[61]

Moreover, during the pandemic, transnational businesses faced steep declines in their revenue and/or production output – proportional to their dependence on the Chinese market. The Seattle-based coffee giant Starbucks, for instance, closed more than half of its 4,292 stores in China[62] – its second-biggest market after the US – with an anticipated financial hit of US$ 915 million.[63] In addition, US automaker Ford halted production at its multiple sites in China, which would otherwise cumulatively produce about half a million vehicles per year, i.e. an average of 9,400 cars per week.[64] As a result, the pandemic expectedly pushed companies to either diversify their investments or alter their supply chains. For instance, the American tech giant Apple – with roughly 70 percent of its revenue generated from products manufactured in China,[65] amidst the pandemic reportedly began looking for alternate suppliers. Apple even sought to reduce its dependence on Foxconn – a Taiwan-based company with a network of facilities in China that produce electronic components for Apple, in order to offset "any expected production loss".[66] The announcement by Apple chief executive Tim Cook came amidst much furore over Apple's announced reduction in production rates and the improbability of reaching its sales targets for 2020 – which had led to the wiping-out of US$ 34 billion in its share market value.[67]

Apart from an emerging international consciousness over their dependencies on China, across the American political spectrum as well, there emerged an increased sympathy for Trump's prescription for reducing the US' interconnectedness with China. Most notably, such a policy acclimatisation at the hands of the coronavirus pandemic became central to the consolidation of a new American domestic political consensus on confronting China.

Fractured American bipartisanship on foreign policy

Traditionally in the US, foreign policy matters enjoyed a sense of bipartisan consensus. Most notably, through the Cold War, Republicans and Democrats – despite being faced with occasional partisan headwinds like the Watergate Scandal, helped in holding the course steady on the US' international relations. Commentators often attribute the same to the early days of the Cold War, when then-Michigan Senator Arthur Vandenberg worked with the Harry Truman administration to forge a bipartisan and bicameral consensus on Capitol Hill with respect to the Truman Doctrine, the Marshall

Plan, and the North Atlantic Treaty Organization. Although this bipartisanship evolved over time under the broader policy of containment of the Soviet Union, that consensus stood the test of time under the Vandenberg dictum of stopping "partisan politics at the water's edge".[68]

In the post–Cold War era, this bipartisan consensus continued to animate American foreign policy. With no near-peer competitor in sight, post–Cold War US foreign policy stemmed from policymakers' commitment to promote liberal *Wilsonian* values, sustain an 'Open Door' global economic system, continue to underwrite the security of allied and partner nations, and sustain high military spending. In the specific case of China, as discussed, a bipartisan liberal internationalist interest to tap into China's market potentialities even helped dampen the US' overt focus on China's record on civil liberties in the post-Tiananmen era.

Over time, however, US bipartisanship on foreign policy came under considerable strain. The hegemonic excesses and follies of an American foreign policy geared towards sustaining US primacy was laid bare with the terror attacks of 11 September 2001 and the protracted military operations that followed. With little strategic gain resulting from the US' regime-change agenda in Iraq and Afghanistan, military adventurism in the post-9/11 timeline proved to be an expensive affair with the US spending nearly US$ 6.5 trillion.[69] In addition, compounded by the financial crisis of 2007–08, a resultant fatigue amongst the American electorate against US activism abroad grew.

Moreover, with Trump's 2016 election victory, the strain on US bipartisanship on foreign policy became more pronounced as the 'America First' worldview forced a conversation on rethinking the US' role in world. During the course of the Trump presidency, that schism mostly played out over Trump's values-bereft conduct of US foreign policy, under its idea of "divorcing" values and foreign policy.[70] In addition, in times of heightened political polarisation, either sides of the American political spectrum have shifted into their respective populist corners. This only further strained the bipartisan consensus on foreign policy. For instance, with the Democratic Party's shift further to the left, the Trump administration's realpolitik 'America First' conduct of US foreign policy spurred the Democratic opposition to espouse an overt focus on furthering *Wilsonian* values through US foreign policy.

However, some erstwhile tenets continue to enjoy bipartisan support. For instance, Republicans and Democrats continue to mostly support high defence spending. In 2019, amidst stalled legislations and polarised debates, Republicans and Democrats managed to come together to finalise federal spending and avert a government shutdown. Both sides hailed another year of high US defence spending, i.e. US$ 738 billion for FY2020.[71] This tenet of traditional US foreign and security policy – which construes high US defence spending as a silver bullet for sustaining American primacy amidst its relative decline in an increasingly multipolar world – has thus continued to enjoy American bipartisan support.

Furthermore, some new areas of convergence are indeed emerging between the conservative nationalist Right and the progressive Left. For instance, both sides are seemingly congruent in their opposition to the US engaging in "endless wars". Although this apparent convergence has not overseen complete actualisation in terms of it translating into concrete legislative actions on the executive's use of force and its ability to deploy US troops abroad without Congressional authorisation, such nascent avenues of a new American bipartisanship on foreign policy are apparent.

Emergence of a new consensus on China

Another such nascent point of convergence between Republicans and Democrats is underpinned by the rising anti-China sentiment amongst the American electorate. As per a Pew poll from March 2020,[72] nearly two-thirds of respondents said that "they have an unfavorable view of China, the most negative rating for the country since the [Pew Research] Center began asking the question in 2005, and up nearly 20 percentage points since the start of the Trump administration". About nine-in-ten US adults identified "China's power and influence" as a threat – with 62 percent recognising it is "a *major* threat". While the total portion that views China as a threat on that question did not change significantly since 2018, the percentage share of those perceiving it as "a *major* threat" increased by 14 percentage points since 2018.

Moreover, as per another Pew poll,[73] Americans' unfavourable views on China rose from 60–26 in 2019 to 66–26 in March 2020. On a partisan scale, Republicans (and leaning Republicans) have consistently held a more unfavourable opinion of China over Democrats (and leaning Democrats). In the Trump years, however, the percentage share for both increased: for Republicans, from 56 percent in 2017 to 72 percent in 2020, and for Democrats, from 41 percent in 2017 to 62 percent in 2020.

In terms of policy, signs of an emergent consensus against China have been apparent. On the Trump administration's centrepiece course of action against China, i.e. levying punitive tariffs on imports from China, Democrats were quick to lend their support. For instance, on the Trump administration's initial round of tariffs in 2018, then-House Minority Leader Nancy Pelosi said, "Beijing's regulatory barriers, localization requirements, labor abuses, anticompetitive 'Made in China 2025' policy and many other unfair trade practices require a full and comprehensive response. The tariffs announced today should be used as a leverage point to negotiate more fair and open trade for U.S. products in China".[74] Notably, even with the Democrats winning control of the House of Representative with their historic win-margins in the 2018 midterms, the support for Trump's confrontational posture against China did not wane.

In February 2020, for instance, after assuming office as House Speaker, Nancy Pelosi led the largest congressional delegation to the NATO Parliamentary Assembly and Munich Security Conference to assuage European concerns amidst fraying transatlantic ties under Trump. However, on urging partner nations to refrain from allowing Huawei to enter their telecommunications market, the delegation of House Democrats was one with Trump. Speaking in Brussels, Pelosi advised allies against using Huawei technology, saying it would mean "selling the privacy of your people down the river". She added, "It's a People's Liberation Army initiative using reversed engineering from Western technology. . . . So of course it's going to be cheaper to put on the market. And if it's cheaper, then they get the market share and then they [China] bring in their autocracy of lack of privacy".[75] Furthermore, with regard to the US' 'special relationship' with the UK, Trump and Pelosi sparred on the former's support for the UK's turn for greater populism. But on the UK's decision to permit Huawei in its 5G trails, Congressional Democrats joined the Trump administration in not only criticising the decision but also raising pressure by suggesting the cessation of US–UK intelligence sharing.[76]

At home, the Democrat-led House even passed a measure to ban the federal government from buying equipment from entities that are deemed national security threats (like Huawei) and sought to create a US$ 1 billion programme to reduce rural networks' dependence on Huawei equipment and replace it with products that are deemed to be relatively "secure".[77] In mid-2020, the Democrat-led House of Representatives and the Republican-led US Senate oversaw introduction of a set of bipartisan legislations – Defending America's 5G Future Act,[78] to ground the Trump administration's actions on Huawei in American law, so as to avoid any prospect of expediency as witnessed in case of the Trump administration back-peddling on ZTE. Lastly, in a sign of consolidating bipartisan will against Chinese unfair commercial practises, the Democrat-controlled House was also set to consider Senate Republicans' legislation on removing rogue Chinese firms from US stock exchanges.[79]

On other foreign policy matters like the US' commitment to allied and partner nations, Democrats in the House sought to dampen Trump's transactional and/or 'America First' inclination via tabling stop-gap legislations to limit Trump's foreign policy powers, like with necessitating Congressional approval for prospective withdrawals from critical alliances. However, on the issue of reinstating focus on China's human rights record, Democrats successfully sought a path to do both – elevate their role in US foreign policy decision-making by contrasting their worldview with Trump's values-bereft conduct *and* complement Trump's policy of sustained confrontation with China.

Some notable examples of such legislations in the Democrat-controlled House include the passage of measures mandating the US secretary of state to certify once a year that Hong Kong has enough autonomy to keep its special US trading consideration, barring the sale of items such as tear gas and rubber bullets to Hong Kong police and requiring China to permit the opening of an American consulate in Tibet before any new Chinese consulate can open in the US.[80]

In addition, House Democrats supported Trump's efforts to further US support for Taiwan. Following Trump's signing of the Taiwan Travel Act – which permits high-level Taiwanese officials to visit the US in their official capacities – House Democrats passed resolutions reaffirming the US' commitment to Taiwan and with the Taiwan Assurance Act of 2019 urged the Trump administration to conduct "regular sales" of defence articles to Taiwan.[81]

However, beyond complementing Trump's approach, there was also an effort to project an American foreign policy that ought to be. For instance, amidst the coronavirus pandemic, Democrats distanced themselves from Trump's insistence on referring to the novel coronavirus as 'Wuhan virus' or 'Chinese virus', and his campaign's effort to deem America to be "under attack – not just by an invisible virus, but by the Chinese".[82] They also were critical of Trump rationalising his unilateralist impulse on terminating the US' relationship with the World Health Organization (WHO) by claiming China to have "total control over the World Health Organization".[83] Similarly, in view of revelations made by Trump's former National Security Advisor John Bolton over Trump's ambivalence towards China's human rights record, Democrats once again raised their apprehensions and pushed back against Bolton's revelation of Trump slow-walking calls to sanction Chinese officials involved in the internment of Uyghur Muslims.

Hence, in this specific context, the US Congress – and the Democrat-led House of Representatives in particular (albeit with crucial support from the Republican-controlled Senate) actively sought to shape the Trump administration's approach. For instance, by mid-2020, the Capitol Hill passed the Uyghur Human Rights Policy Act of 2020 to unanimously mandate the executive branch to impose sanctions against Chinese officials responsible for the internment of Uyghurs in Xinjiang[84] and passed the Hong Kong Autonomy Act to mandate sanctions against Chinese officials/entities that were involved in the imposition of Beijing's crippling national security law in Hong Kong.[85]

Thus, with the emergence of this nascent US bipartisanship on China, the coronavirus pandemic will possibly be looked back upon as a watershed moment in the evolving US–China strategic rivalry.

Notes

1 Edmund Andrews, "U.S. Adds Tariffs on Chinese Tires", *The New York Times*, 11 September 2009, last accessed 17 September 2020, www.nytimes.com/2009/09/12/business/global/12tires.html

2 Molly Moorhead, "Ad Says Obama Defied China to Save Tire-making Jobs at Home, Romney Criticized", *Politifact*, 01 October 2012, last accessed 17 September 2020, www.politifact.com/factchecks/2012/oct/01/barack-obama/ad-says-obama-defied-china-save-tire-making-jobs-h/
3 Barack Obama, "Remarks by the President in State of the Union Address", *The Barack Obama Administration – The White House Archives*, 24 January 2012, last accessed 17 September 2020, https://obamawhitehouse.archives.gov/the-press-office/2012/01/24/remarks-president-state-union-address
4 Matthew Yglesias, "The Democratic Party's Down-ballot Collapse, Explained", *Vox*, 10 January 2017, last accessed 17 September 2020, www.vox.com/policy-and-politics/2017/1/10/14211994/obama-democrats-downballot
5 *Quoted in* Andy Barr, "The GOP's No-compromise Pledge", *Politico*, 28 October 2010, last accessed 17 September 2020, www.politico.com/story/2010/10/the-gops-no-compromise-pledge-044311
6 Harold Pachios, "Let's Look at China's Role in the Iran Nuclear Deal", *The Hill*, 21 August 2015, last accessed 17 September 2020, https://thehill.com/blogs/congress-blog/foreign-policy/251657-lets-look-at-chinas-role-in-the-iran-nuclear-deal
7 Tanya Somanader, "President Obama: The United States Formally Enters the Paris Agreement", *The Barack Obama Administration – The White House Archives*, 03 September 2016, last accessed 17 September 2020, https://obamawhitehouse.archives.gov/blog/2016/09/03/president-obama-united-states-formally-enters-paris-agreement
8 Patrick Gillespie, "Obama Got Tough on China. It Cost U.S. Jobs and Raised Prices", *CNN*, 03 January 2017, last accessed 17 September 2020, https://money.cnn.com/2017/01/03/news/economy/obama-china-tire-tariff/index.html
9 Xi Jingping, "Remarks by President Obama and President Xi of the People's Republic of China in Joint Press Conference", *The Barack Obama Administration – The White House Archives*, 25 September 2015, last accessed 16 September 2020, https://obamawhitehouse.archives.gov/the-press-office/2015/09/25/remarks-president-obama-and-president-xi-peoples-republic-china-joint
10 BBC report, "Donald Trump's Top Four 'Deals' for Foreign Policy", *BBC*, 27 April 2016, last accessed 17 September 2020, www.bbc.com/news/election-us-2016-36154988
11 Donald Trump, "Remarks by President Trump to the 74th Session of the United Nations General Assembly", *The Donald Trump Administration – The White House Archives*, 25 September 2019, last accessed 17 September 2020, www.whitehouse.gov/briefings-statements/remarks-president-trump-74th-session-united-nations-general-assembly/
12 *Quoted in* PBS transcript, "Steve Bannon – First Interview with the Former Trump Adviser", *Frontline – Public Broadcasting Service*, 17 March 2019, last accessed 16 September 2020, www.pbs.org/wgbh/frontline/interview/steve-bannon-3/
13 *Quoted in* PBS transcript, "Steve Bannon – First Interview with the Former Trump Adviser", *Frontline – Public Broadcasting Service*, 17 March 2019, last accessed 16 September 2020, www.pbs.org/wgbh/frontline/interview/steve-bannon-3/
14 Kashish Parpiani, "Correlation between American Policy Schizophrenia with China, and Indian Funambulism on Integrating with the U.S.-led Liberal Order", *Rising Powers Quarterly*, Volume 4, Issue 1 (The Fate of the Liberal International Order and Rising Powers), Aug. 2019, pp. 41–55.
15 Robert Zoellick, "Whither China: from Membership to Responsibility? – Robert B. Zoellick, Deputy Secretary of State – Remarks to National Committee on U.S.–China Relations", *The United States Department of State – Archives*,

21 September 2005, last accessed on 17 September 2020, https://2001-2009.state.gov/s/d/former/zoellick/rem/53682.htm

16 BBC report, "Senate Approves China Trade Deal", *BBC*, 19 September 2000, last accessed 17 September 2020, http://news.bbc.co.uk/1/hi/world/americas/932450.stm

17 Kristy Alfredson, "Bush Backs China's WTO Entry Despite Standoff", *CNN*, 06 April 2001, last accessed 17 September 2020, http://edition.cnn.com/2001/WORLD/asiapcf/east/04/05/china.WTO

18 Paul Blustein, "The Untold Story of How George W. Bush Lost China", *Foreign Policy*, 02 October 2019, last accessed 17 September 2020, https://foreignpolicy.com/2019/10/04/the-untold-story-of-how-george-w-bush-lost-china/

19 *Quoted in* Kristy Alfredson, "Bush Backs China's WTO Entry Despite Standoff", *CNN*, 06 April 2001, last accessed 17 September 2020, http://edition.cnn.com/2001/WORLD/asiapcf/east/04/05/china.WTO

20 Federica Cocco, "Most US Manufacturing Jobs Lost to Technology, Not Trade", *The Financial Times*, 02 December 2016, last accessed 17 September 2020, www.ft.com/content/dec677c0-b7e6-11e6-ba85-95d1533d9a62

21 Treasury Department press release, "Benefits of Trade with China", *The US Department of the Treasury – Archives*, 24 May 2007, last accessed 17 September 2020, www.treasury.gov/press-center/press-releases/Pages/hp423.aspx

22 Pew poll, "Top Voting Issues in 2016 Election", *Pew Research Center*, 07 July 2016, last accessed 17 September 2020, www.pewresearch.org/politics/2016/07/07/4-top-voting-issues-in-2016-election/

23 Bruce Drake and Carroll Doherty, "Key Findings on How Americans View the U.S. Role in the World", *Pew Research Center*, 05 May 2016, last accessed 17 September2020,www.pewresearch.org/fact-tank/2016/05/05/key-findings-on-how-americans-view-the-u-s-role-in-the-world/

24 NSS 2017, "National Security Strategy of the United States of America", *The Donald Trump Administration – The White House Archives*, December 2017, pp. 25–7, last accessed 17 September 2020, www.whitehouse.gov/wp-content/uploads/2017/12/NSS-Final-12-18-2017-0905.pdf

25 Fox Business report, "Trump Warns That $800 Billion Trade Deficit Is 'Unacceptable'", *Fox Business*, 15 November 2017, last accessed 17 September 2020, www.foxbusiness.com/politics/trump-warns-that-800-billion-trade-deficit-is-unacceptable

26 White House factsheet, "President Donald J. Trump Is Promoting Free, Fair, and Reciprocal Trade", *The Donald Trump Administration – The White House Archives*, 30 January 2018, last accessed 17 September 2020, www.whitehouse.gov/briefings-statements/president-donald-j-trump-promoting-free-fair-reciprocal-trade/

27 Donald Trump, "Remarks by President Trump to the 74th Session of the United Nations General Assembly", *The Donald Trump Administration – The White House Archives*, 25 September 2019, last accessed 17 September 2020, www.whitehouse.gov/briefings-statements/remarks-president-trump-74th-session-united-nations-general-assembly/

28 Amanda Macias, "WTO Says U.S. Tariffs on Chinese Goods Violated International Trade Rules", *CNBC*, 15 September 2020, last accessed 17 September 2020, www.cnbc.com/2020/09/15/wto-says-us-tariffs-on-chinese-goods-violated-international-trade-rules.html

29 *Quoted in* Marc-William Palen, "Trump Says Trade Wars Are Easy to Win. Here's What History Tells Us About a Possible U.S.–China Conflict", *NBC News*, 16 April 2018, last accessed 17 September 2020, www.cnbc.com/2018/03/02/trump-trade-wars-are-good-and-easy-to-win.html

30 Reuters report, "Timeline: Key Dates in the U.S.–China Trade War", *Reuters*, 15 January 2020, last accessed 17 September 2020, www.reuters.com/article/us-usa-trade-china-timeline/timeline-key-dates-in-the-u-s-china-trade-war-idUSKBN1ZE1AA
31 BBC report, "A Quick Guide to the US–China Trade War", *BBC*, 16 January 2020, last accessed 17 September 2020, www.bbc.com/news/business-45899310
32 AFBF report, "China Trade Standstill More Bad News for Farmers", *American Farm Bureau Federation*, 05 August 2019, last accessed 17 September 2020, www.fb.org/newsroom/china-trade-standstill-more-bad-news-for-farmers
33 Stuart Anderson, "Trump Tariff Aid to Farmers Cost More Than U.S. Nuclear Forces", *Forbes*, 21 January 2020, last accessed 17 September 2020, www.forbes.com/sites/stuartanderson/2020/01/21/trump-tariff-aid-to-farmers-cost-more-than-us-nuclear-forces/#1acfd24c6c50
34 White House factsheet, "President Donald J. Trump Is Signing a Landmark Phase One Trade Agreement with China", *The Donald Trump Administration – The White House Archives*, 15 January 2020, last accessed 17 September 2020, www.whitehouse.gov/briefings-statements/president-donald-j-trump-signing-landmark-phase-one-trade-agreement-china/
35 Stuart Anderson, "Trump Tariff Aid to Farmers Cost More Than U.S. Nuclear Forces", *Forbes*, 21 January 2020, last accessed 17 September 2020, www.forbes.com/sites/stuartanderson/2020/01/21/trump-tariff-aid-to-farmers-cost-more-than-us-nuclear-forces/#1acfd24c6c50
36 France 24 report, "US and China Sign 'Phase One' Trade Deal, Ending Dramatic 18-Month Tit-for-Tat", *France 24 News*, 15 January 2020, last accessed 17 September 2020, www.france24.com/en/20200115-us-and-china-sign-phase-one-trade-deal-aimed-at-easing-tensions
37 Laura Zhou, "Trump Signs off on Plan to Allow US Navy More Freedom to Patrol in South China Sea, Report Says", *South China Morning Post*, 22 July 2017, last accessed 17 September 2020, www.scmp.com/news/china/diplomacy-defence/article/2103718/trump-signs-plan-allow-us-navy-more-freedom-patrol
38 Laura Zhou, "Trump Signs off on Plan to Allow US Navy More Freedom to Patrol in South China Sea, Report Says", *South China Morning Post*, 22 July 2017, last accessed 17 September 2020, www.scmp.com/news/china/diplomacy-defence/article/2103718/trump-signs-plan-allow-us-navy-more-freedom-patrol
39 Idrees Ali and David Brunnstrom, "U.S. Warship Drill Meant to Defy China's Claim over Artificial Island: Officials", *Reuters*, 25 May 2017, last accessed 17 September 2020, www.reuters.com/article/us-usa-southchinasea-navy-idUSKBN18K353
40 Ralph Jennings, "Why Trump Has Gotten Extra Tough in Monitoring China at Sea", *VOA News*, 21 June 2019, last accessed 17 September 2020, www.voanews.com/east-asia/why-trump-has-gotten-extra-tough-monitoring-china-sea
41 David Larter, "In Challenging China's Claims in the South China Sea, the US Navy Is Getting more Assertive", *Defense News*, 05 February 2020, last accessed 17 September 2020, www.defensenews.com/naval/2020/02/05/in-challenging-chinas-claims-in-the-south-china-sea-the-us-navy-is-getting-more-assertive/
42 David Larter, "In Challenging China's Claims in the South China Sea, the US Navy Is Getting more Assertive", *Defense News*, 05 February 2020, last accessed 17 September 2020, www.defensenews.com/naval/2020/02/05/in-challenging-chinas-claims-in-the-south-china-sea-the-us-navy-is-getting-more-assertive/
43 *Quoted in* Ralph Jennings, "Why Trump Has Gotten Extra Tough in Monitoring China at Sea", *VOA News*, 21 June 2019, last accessed 17 September 2020, www.voanews.com/east-asia/why-trump-has-gotten-extra-tough-monitoring-china-sea

44 Brad Lendon, "US Navy Stages Back-to-back Challenges to Beijing's South China Sea Claims", *CNN*, 30 April 2020, 17 September 2020, https://edition.cnn.com/2020/04/30/asia/south-china-sea-us-navy-freedom-of-navigation-intl-hnk/index.html
45 David Larter, "In Challenging China's Claims in the South China Sea, the US Navy Is Getting More Assertive", *Defense News*, 05 February 2020, last accessed 17 September 2020, www.defensenews.com/naval/2020/02/05/in-challenging-chinas-claims-in-the-south-china-sea-the-us-navy-is-getting-more-assertive/
46 *Quoted in* David Larter, "In Challenging China's Claims in the South China Sea, the US Navy Is Getting More Assertive", *Defense News*, 05 February 2020, last accessed 17 September 2020, www.defensenews.com/naval/2020/02/05/in-challenging-chinas-claims-in-the-south-china-sea-the-us-navy-is-getting-more-assertive/
47 Ankit Panda, "South China Sea: Fourth US FONOP in Five Months Suggests a New Operational Rhythm", *The Diplomat*, 12 October 2017, last accessed 17 September 2020, https://thediplomat.com/2017/10/south-china-sea-fourth-us-fonop-in-five-months-suggests-a-new-operational-rhythm/
48 Stephanie Nebehay, "In a First, China Knocks U.S. from Top Spot in Global Patent Race", *Reuters*, 07 April 2020, last accessed 17 September 2020, www.reuters.com/article/us-usa-china-patents/in-a-first-china-knocks-us-from-top-spot-in-global-patent-race-idUSKBN21P1P9
49 WIPO press release, "China Becomes Top Filer of International Patents in 2019 Amid Robust Growth for WIPO's IP Services, Treaties and Finances", *World Intellectual Property Organization*, 07 April 2020, last accessed 17 September 2020, www.wipo.int/pressroom/en/articles/2020/article_0005.html
50 Zen Soo and Sarah Dai, "How Did Huawei Fall Foul of the US Government and Find Itself at the Epicentre of a New Tech War?", *South China Morning Post*, 06 April 2020, last accessed 17 September 2020, www.scmp.com/techenterprises/article/3078451/how-did-huawei-fall-foul-us-government-and-find-itself-epicentre
51 Justice Department press release, "Chinese Telecommunications Conglomerate Huawei and Subsidiaries Charged in Racketeering Conspiracy and Conspiracy to Steal Trade Secrets", *The US Department of Justice*, 13 February 2020, last accessed 17 September 2020, www.justice.gov/opa/pr/chinese-telecommunications-conglomerate-huawei-and-subsidiaries-charged-racketeering
52 Maggie Miller, "FCC Votes to Bar Use of Its Funds to Purchase Huawei, ZTE Equipment", *The Hill*, 22 November 2019, last accessed 17 September 2020, https://thehill.com/policy/technology/471685-fcc-votes-to-bar-use-of-its-funds-to-purchase-huawei-zte-equipment
53 *Quoted in* NYT report, "Really? Is the White House Proposing to Buy Ericsson or Nokia?", *The New York Times*, 07 February 2020, last accessed 17 September 2020, www.nytimes.com/2020/02/07/business/dealbook/bill-barr-huawei-nokia-ericsson.html
54 Pew poll, "U.S. Views of China Increasingly Negative Amid Coronavirus Outbreak", *Pew Research Center*, 21 April 2020, last accessed 17 September 2020, www.pewresearch.org/global/2020/04/21/u-s-views-of-china-increasingly-negative-amid-coronavirus-outbreak/
55 Jodi Xu Klein, "Donald Trump Gets His Way as US Lifts ZTE Ban, Allowing Company to Continue Despite Senators' Opposition", *South China Morning Post*, 14 July 2018, last accessed 17 September 2020, www.scmp.com/news/china/economy/article/2155238/us-lifts-ban-american-companies-selling-parts-chinas-zte
56 Ana Swanson, "U.S. Delivers Another Blow to Huawei with New Tech Restrictions", *The New York Times*, 15 May 2020, last accessed 17 September 2020, www.nytimes.com/2020/05/15/business/economy/commerce-department-huawei.html

57 David Shepardson, "EXCLUSIVE-Trump to Block U.S. Downloads of TikTok, WeChat on Sunday-Officials", *Reuters*, 18 September 2020, last accessed 18 September 2020, https://in.reuters.com/article/usa-tiktok-ban/exclusive-trump-to-block-u-s-downloads-of-tiktok-wechat-on-sunday-officials-idUSL1N2GF039
58 IBISworld report, "Tire Manufacturing Industry in China – Market Research Report", *IBISWorld*, 05 March 2020, last accessed 17 September 2020, www.ibisworld.com/china/market-research-reports/tire-manufacturing-industry/
59 Rory Horner, "The World Needs Pharmaceuticals from China and India to Beat Coronavirus", *The Conversation*, 25 May 2020, last accessed 17 September 2020, https://theconversation.com/the-world-needs-pharmaceuticals-from-china-and-india-to-beat-coronavirus-138388
60 Alexandra Stevenson, "Coronavirus Outbreak Tests World's Dependence on China", *The New York Times*, 29 January 2020, last accessed 17 September 2020, www.nytimes.com/2020/01/29/business/china-coronavirus-economy.html
61 Alexandra Stevenson, "Coronavirus Outbreak Tests World's Dependence on China", *The New York Times*, 29 January 2020, last accessed 17 September 2020, www.nytimes.com/2020/01/29/business/china-coronavirus-economy.html
62 Heather Haddon and Micah Maidenberg, "Starbucks Closes Half of China Stores Amid Coronavirus Outbreak", *The Wall Street Journal*, 28 January 2020, last accessed 17 September 2020, www.wsj.com/articles/starbucks-closed-half-of-china-stores-amid-coronavirus-outbreak-11580247048
63 Megan Campbell, "Starbucks Estimates $915M Loss from Coronavirus Impact", *Business Journal*, 30 April 2020, last accessed 17 September 2020, www.bizjournals.com/seattle/news/2020/04/30/starbucks-estimates-915m-loss-from-coronavirus.html
64 Alexandra Stevenson, "Coronavirus Outbreak Tests World's Dependence on China", *The New York Times*, 29 January 2020, last accessed 17 September 2020, www.nytimes.com/2020/01/29/business/china-coronavirus-economy.html
65 Yoko Kubota, "Coronavirus Exposes Businesses' Dependency on China", *The Wall Street Journal*, 18 February 2020, last accessed 17 September 2020, www.wsj.com/articles/coronavirus-exposes-businesses-dependency-on-china-11582035993
66 Bloomberg report, "CEO Tim Cook Says Apple Working to Mitigate Impact of China Coronavirus", *South China Morning Post*, 29 January 2020, last accessed 17 September 2020, www.scmp.com/tech/big-tech/article/3047977/apple-holiday-results-top-estimates-rebounding-iphone-demand
67 Julia Horowitz, "Apple's Coronavirus Warning Just Shaved $34 Billion off Its Stock Market Value", *CNN*, 18 February 2020, last accessed 17 September 2020, https://edition.cnn.com/2020/02/18/investing/premarket-stocks-trading/index.html
68 Senate archive, "Arthur H. Vandenberg", *The US Senate*, n.d., last accessed 17 September 2020, www.senate.gov/artandhistory/art/artifact/Painting_32_00042.htm
69 Amanda Macias, "America Has Spent $6.4 Trillion on Wars in the Middle East and Asia Since 2001, a New Study Says", *CNBC*, 20 November 2019, last accessed 17 September 2020, www.cnbc.com/2019/11/20/us-spent-6point4-trillion-on-middle-east-wars-since-2001-study.html
70 Julian Borger, "Rex Tillerson: 'America First' Means Divorcing Our Policy from Our Values", *The Guardian*, 04 May 2017, last accessed 17 September 2020, www.theguardian.com/us-news/2017/may/03/rex-tillerson-america-first-speech-trump-policy
71 Joe Gould, "Pentagon Finally Gets Its 2020 Budget from Congress", *Defense News*, 19 December 2019, last accessed 17 September 2020, www.defensenews.com/congress/2019/12/19/pentagon-finally-gets-its-2020-budget-from-congress/
72 Kat Devlin, Laura Silver, and Christine Huang, "U.S. Views of China Increasingly Negative Amid Coronavirus Outbreak", *Pew Research Center*, 21 April

2020, last accessed 17 September 2020, www.pewresearch.org/global/2020/04/21/u-s-views-of-china-increasingly-negative-amid-coronavirus-outbreak/
73 Laura Silver, Kat Devlin, and Christine Huang, "U.S. Views of China Turn Sharply Negative Amid Trade Tensions", *Pew Research Center*, 13 August 2019, last accessed 17 September 2020, www.pewresearch.org/global/2019/08/13/u-s-views-of-china-turn-sharply-negative-amid-trade-tensions/
74 Rep. Nancy Pelosi press release, "Pelosi Statement on Trump Administration's New Tariffs on China", *Congresswoman Nancy Pelosi*, 22 March 2018, last accessed 17 September 2020, https://pelosi.house.gov/news/press-releases/pelosi-statement-on-trump-administration-s-new-tariffs-on-china
75 *Quoted in* Joe Gould, "Congress' Foreign Policy Flex in Europe", *Defense News*, 22 February 2019, last accessed 17 September 2020, www.defensenews.com/congress/2019/02/22/congress-foreign-policy-flex-in-europe/
76 Maggie Miller, "Lawmakers Warn US, UK Intel Sharing at Risk after Huawei Decision", *The Hill*, 28 January 2020, last accessed 17 September 2020, https://thehill.com/policy/cybersecurity/480353-lawmakers-warn-us-uk-intel-sharing-at-risk-after-huawei-decision
77 Emily Birnbaum, "House Passes Legislation Banning Government from Buying Huawei Equipment", *The Hill*, 16 December 2019, last accessed 17 September 2020, https:/thehill.com/policy/technology/474790-house-passes-legislation-banning-government-from-buying-huawei-equipment
78 Reuters report, "Bills Targeting China's Huawei Introduced in Congress", *Reuters*, 16 July 2019, last accessed 17 September 2020, www.reuters.com/article/us-usa-china-huawei-congress/bills-targeting-chinas-huawei-introduced-in-congress-idUSKCN1UB1X3
79 Cristina Marcos, "Pelosi Says House Is Looking at Bill That Could Delist Some Chinese Companies from US Stock Exchanges", *The Hill*, 21 May 2020, last accessed 17 September 2020, https://thehill.com/homenews/house/499005-pelosi-says-house-is-looking-at-bill-that-could-delist-some-chinese-companies
80 Paul Eckert, "Bill to Support Tibet Wins Strong Approval in US House of Representatives", *Radio Free Asia*, 28 January 2020, last accessed 17 September 2020, www.rfa.org/english/news/tibet/tibet-congress-01282020181342.html
81 Reuters report, "U.S. House Passes Pro-Taiwan Bills, Amid Trade Tensions with China", *Reuters*, 08 May 2019, last accessed 17 September 2020, www.reuters.com/article/us-usa-china-taiwan-congress/us-house-passes-pro-taiwan-bills-amid-trade-tensions-with-china-idUSKCN1SE00T
82 Trump Campaign press release, "In Midst of Coronavirus Crisis, Joe Biden Defends China and Parrots Communist Party Propaganda", *Trump-Pence 2020 Campaign Website*, 18 March 2020, last accessed 17 September 2020, www.donaldjtrump.com/media/in-midst-of-coronavirus-crisis-joe-biden-defends-china-and-parrots-communist-party-propaganda/
83 *Quoted in* BBC report, "Coronavirus: Trump Terminates US Relationship with WHO", *BBC*, 30 May 2020, last accessed 17 September 2020, www.bbc.com/news/world-us-canada-52857413
84 Reuters report, "U.S. Lawmakers Say Uighur Bill Supports China's Muslims", *Reuters*, 02 June 2020, last accessed 17 September 2020, https://in.reuters.com/article/usa-china-xinjiang/u-s-lawmakers-say-uighur-bill-supports-chinas-muslims-idINKBN2392EM
85 Kevin Breuninger and Amanda Macias, "China Vows Retaliation after Trump Slaps Sanctions on It for Interference in Hong Kong", *CNBC*, 14 July 2020, last accessed 17 September 2020, www.cnbc.com/2020/07/14/trump-signed-law-slapping-sanctions-on-china-for-interference-in-hong-kong.html

3

TRUMP AND THE INDO-PACIFIC

The Trump administration complemented its sustained confrontational agenda on China with a region-wide strategy. Building upon the predecessor administration's Pivot to Asia policy, the Trump administration sought to expand its purview of the region with the Indo-Pacific construct. In recognising the interlinked destinies of the Pacific and Indian oceans, the Trump administration underscored its outlook to the expanded region as being towards securing a "Free and Open Indo-Pacific". Beyond its focus on rising Chinese influence –in terms of either its territorial aggrandisement in the South China Sea or its cultivation of strategic vassals via its policy of 'debt-trap' diplomacy – the US adopted that normative dimension of ensuring an Indo-Pacific that is centred on a rules-based order – imbued with shared values and principles.

President Trump's iteration of the same first came at the 2017 APEC CEO Summit in Da Nang, Vietnam. In expressing his honour to underline the US' vision for the Indo-Pacific, he purported it to be as "a place where sovereign and independent nations, with diverse cultures and many different dreams, can all prosper side-by-side, and thrive in freedom and in peace".[1] In line with this outlook, the administration also renamed the US Pacific Command in Hawaii to the US Indo-Pacific Command. In defining the Hawaii-based US Combatant Command to be the US' "primary combatant command", then-Secretary of Defense James Mattis construed the US' outlook: "All nations large and small are essential to the region, in order to sustain stability in ocean areas critical to global peace".[2]

Beyond the rhetorical flourish, in at least four critical documents released by the Trump administration, the US' focus on the Indo-Pacific was underscored. Starting with its maiden National Security Strategy (NSS), the administration announced the return of the era of "great power competition". In stating that "a geopolitical competition between free and repressive visions of world order is taking place in the Indo-Pacific region", the NSS outlined the region-wide challenge posed by China:

> Although the United States seeks to continue to cooperate with China, China is using economic inducements and penalties, influence oper-

> ations, and implied military threats to persuade other states to heed its political and security agenda. China's infrastructure investments and trade strategies reinforce its geopolitical aspirations. Its efforts to build and militarize outposts in the South China Sea endanger the free flow of trade, threaten the sovereignty of other nations, and undermine regional stability. China has mounted a rapid military modernization campaign designed to limit U.S. access to the region and provide China a freer hand there. China presents its ambitions as mutually beneficial, but Chinese dominance risks diminishing the sovereignty of many states in the Indo-Pacific.[3]

Thereafter, in the 2019 US State Department release of the "A Free and Open Indo-Pacific – Advancing a Shared Vision" policy document, the administration emphasised the normative aims of its strategy. It stated: "The U.S. vision for the Indo-Pacific excludes no nation. We do not ask countries to choose between one partner or another. Instead, we ask that they uphold the core principles of the regional order at a time when these principles are under renewed threat".[4] This document emerged not only as the most definitive account of the Trump administration's implementation of its Indo-Pacific strategy but also as the administration's most comprehensive iteration of its aims in the region. For instance, in picking up on the predecessor administration's assertion of the US being "a Pacific power", the document recounts Trump's commitment to the region: "We [the US] have been friends, partners, and allies in the Indo-Pacific for a long, long time, and we will be friends, partners, and allies for a long time to come".[5]

In addition, the document underscored the Trump administration's multifaceted approach to the region as going beyond merely shoring up traditional US allies and partners in the East Pacific region. In recounting its gains with nascent partners like India, the document highlighted the importance the Trump administration "places on South Asian partners' political autonomy and their role in an inclusive Indo-Pacific vision".[6] Furthermore, in complementing its outreach in the strategic domain, the document underscores the US' economic vision for the region: "We [the US] respect the sovereignty of every nation, and our economic engagement seeks to equip states to resist coercive economic practices, unsustainable debt burdens, and other dangers. We do this by improving market access and competitiveness, facilitating business-to-business ties, and promoting free, fair, and reciprocal trade".[7] This categorical pledge is followed by dedicated sections on the Trump administration's impetus to the region's infrastructure, energy, and digital economy. In positing a constructive vision – which would go beyond the Obama administration's status quoist Pivot/Rebalance to Asia approach, the document notes the Trump administration's development aid and assistance to have been geared towards free-market ideals such as "citizen-responsive governance and light-touch regulation that fosters entrepreneurship and the efficient allocation of capital, and institutions that promote transparency, fairness, and the sanctity of contracts".[8] Finally, in listing the Trump administration's impetus to these domains that go beyond the US' deep-rooted security role in the region, the document closes with noting the progress made by the State Department and USAID since 2017. In asserting the US to have made a dedicated shift of resources to the Indo-Pacific compared to "the last three years" of the Obama administration, the document notes the Trump administration to have disbursed "more than $4.5 billion in foreign assistance to the region" –representing an increase of 25 percent.[9]

Finally, with the National Defense Strategy (NDS) 2018[10] (the first since its 2008 iteration) and the 2019 Indo-Pacific Strategy Report[11] – both released by the US Department of Defense, the administration outlined its strategy to primarily depend on strengthening "alliances and partnerships in the Indo-Pacific to a networked security architecture capable of deterring aggression, maintaining stability, and ensuring free access to common domains".[12]

Trump and US allies in the region

With repeated invocation of the US' security and economic commitments in the Trump administration's national security literature, the criticality of US alliances in the Indo-Pacific was also recognised. Especially, in view of the region being rife with historical rivalries which at times even impede cooperation between two or more US allies. Moreover, unlike the North Atlantic Treaty Organization in Europe, US alliances in the Indo-Pacific are not placed under one integrative ambit. Rather, they are spread out in bilateral arrangements (chiefly with Japan, South Korea, Australia, and the Philippines) with the US to form the US' 'hub & spokes'[13] alliance system in the region. However, despite recognising the importance of this network of bilateral partnerships, the Trump administration's record was mixed.

Relationships weathering the storm

One of the critical partnerships in the region which fared relatively well under Trump is the US' alliance with **Japan**. The nodal US security alliance in the region not only held steady under the Trump presidency, it flourished to emerge as an axiomatic relationship irrespective of political headwinds in the US that denounce American internationalism. The alliance mostly witnessed growth in the Trump years, at the hands of an unusually strong personal dynamic between the American president and Japanese Prime Minister Shinzo Abe. Between Trump's election victory in late 2016 and early 2020, the leaders "met 20 times, played 5 rounds of golf, and had 32 phone calls, at times speaking twice a week".[14]

Although alliances and security partnerships thrive on institutionalised mechanisms, in the case of the US–Japan dynamic – which has marked over

60 years, policy developments have required an occasional push from the chemistry between the nations' top political leadership. Especially so in the case of Trump's presidency, which was not only known for its abhorrence of US commitments towards partner nations but also witnessed a considerable amount of personalisation of US foreign policy. Such an impetus to personality-level ties became imperative from Japan's standpoint, also due to Trump's campaign rhetoric over pulling out US troops (nearly 50,000)[15] from Japan, if Tokyo did not agree to new cost-sharing terms (despite Japan already offsetting nearly 74.5 percent of the total cost).[16] Trump had even suggested that US allies in Northeast Asia (chiefly, Japan and South Korea) could be better off with acquiring their own nuclear weapons to protect themselves against North Korea's brinksmanship.[17]

As a result, following Trump's victory, Abe became the first foreign leader to meet with then-President-elect Trump. Ahead of the meeting which lasted nearly two hours, Abe deemed the purpose of the visit to "build trust".[18] In terms of optics, this meeting set the US–Japan trajectory on a firm footing with adversarial nations possibly taking note of the partnership's resilience irrespective of Trump's criticisms on the campaign trail. Another moment of high symbolism came in early 2017 during the leaders' first summit after Trump's inauguration. At a press conference standing alongside Abe, Trump said, "[T]he United States of America is behind Japan, our great ally, 100 percent".[19]

This personal camaraderie also translated into policy-level convergences, in evident contradiction to Trump's 'America First' outlook. Chiefly, in accommodating Japanese interests under the ambit of US negotiations with North Korea, Trump met families of Japanese citizens abducted by North Korea,[20] mentioned the issue in his 2017 address to the 72nd Session of the United Nations General Assembly,[21] and reportedly even pursued the matter in his meetings with Kim Jong-un in Hanoi, Vietnam.[22] Another notable aspect was a dampening of Trump's rhetoric over Japan "free-riding" on US security guarantees. In a move reflective of considerable continuity with the predecessor Obama administration, Trump's Secretary of Defense James Mattis not only reaffirmed the US' commitment to defend Japan but also clarified that the disputed Senkaku Islands fall under the purview of US commitments. At a press conference with Japanese Defense Minister Tomomi Inada, Mattis said, "I made clear that our long-standing policy on the Senkaku Islands stands – the US will continue to recognise Japanese administration of the islands and as such Article 5 of the US–Japan Security Treaty applies".[23]

On trade, although the Trump administration contemplated a "sweeping move" to levy tariffs on Japanese cars on the basis of national security (much like its steel and aluminium tariffs),[24] trade frictions steadied over time – albeit with the continued imposition of 2.5 percent auto tariffs.[25] In late 2019, the two sides even signed a limited trade deal for increased market-access into Japan – to further gains on Japan already being America's third-largest agricultural export market – and specifically included market-opening commitments on US$ 40 billion worth of digital trade.[26] Although Abe did not exact a commitment from Trump on not raising US tariffs on Japanese autos under the US' Section 232 national security provision, Japan did score a reasonable deal from the standpoint of Abe's domestic political promise of holding the line against American demands for "better-than-TPP access to the Japanese agriculture market".[27]

As for Trump's repeated derision of America's trade deficit with Japan, which was reported to be US$ 58 billion in 2018,[28] an impetus to defence trade managed to soothe tempers. In line with the Trump administration's push for US arms exports under its 'Buy American' policy, the Abe government's decision to increase its purchase orders of US-made platforms like the F-35 fifth-generation Lightning II stealth fighter jet was a notable development. In late 2018, Abe's cabinet approved the increase of Japan's order of the F-35 aircraft from 42 to 147.[29] The additional order – comprised of 63 F-35As (the conventional take-off and landing variant) and 42 F-35Bs (the short take-off and vertical landing (STOVL) variant) – made Japan the biggest international buyer of the coveted fighter jet.[30] This briefly also fed the prospect of Japan moving "from being a customer of the F-35 to a full-fledged member of the industrial base consortium".[31] Moreover, in face of increased pressure from Trump for Japan to increase imports of US weaponry and platforms to correct trade imbalances,[32] Abe's conservative agenda to shed principles of constitutional pacifism towards rebuilding Japan's military gained an apt political cover.

Most importantly, the Trump administration accepted and embraced Tokyo's call for "a free and open Indo-Pacific". In recognising the interlinked destinies of the Pacific and Indian oceans, Abe had long been an advocate for the cultivation of a rules-based order in the region to counter Chinese hegemonic intentions.[33] In marrying its approach to the Indo-Pacific with Abe's goal for a "free and open" region, the Trump administration illustrated its commitment to the US' alliance with Japan.

With its 2016 Defence White Paper and 2017 Foreign Policy White Paper, **Australia** was another regional nation which had been an advocate for those aims in the Indo-Pacific region.[34] Hence, much like the case of US–Japan ties, US–Australia ties too fared well under Trump. For its second state dinner, for instance, the Trump administration hosted Australian Prime Minister Scott Morrison in September 2019. This marked the first US state dinner hosted for an Australian PM since John Howard was received by the George W Bush administration in 2006.[35] In its effort to shore up allies, the Obama administration had identified Australia as a lynchpin of the US' strategy in the Southern Pacific. Cooperating in the Pacific was seen as a matter of natural convergence between the two countries, as Australian troops have participated "alongside U.S. troops in every war the United States has entered since World War I, and the two countries have been bound by a mutual defense agreement – the Australia, New Zealand, and U.S. Security (ANZUS) Treaty – since 1951".[36]

Recognising the strong relevance of this partnership in context of its focus on the Asia-Pacific region, in 2011, the Obama administration had committed to a 2,500-size US Marines presence in Darwin, Australia. Although plans for the same momentarily hit a block over cost-sharing arrangements for housing the US Marines, the Trump administration, in an apparent sign of continuity, fulfilled the Obama-era goal by mid-2019. Although it did so in its renewed context of the Indo-Pacific strategy, the Trump administration built on the 200-person company stationed in Darwin. By spring of 2019, the "eighth iteration of the Marine Rotational Force was 1,700 strong, including infantry Marines, logistics capabilities and aircraft".[37] Following this, another 800 Marines arrived in Australia in mid-2019 – including infantry, logistics and medical personnel. On the occasion of the Trump administration following through on the Obama-era goal, Australian Defence Minister Linda Reynolds said, "This milestone demonstrates the enduring nature of the Australia–U.S. alliance and our deep engagement with the Indo-Pacific region".[38] She added, "The Marine Rotational Force–Darwin improves interoperability between Australian and U.S. defence forces, and enhances our ability to work together with regional partners in the interests of stability and security in the Indo-Pacific".

In 2019, the US and Australia also oversaw the biggest iteration of the Talisman Saber, a biennial joint military exercise "designed to ensure and demonstrate the ability of the two defense forces to work together with the highest levels of interoperability".[39] The same encompassed "over 34,000 personnel, 22 ships and over 200 aircraft".[40] Furthermore, under Trump, Australia and the US witnessed convergences on other matters such as Canberra becoming the third country to join Britain and Bahrain in the US' effort to protect transnational commercial ships from Iranian threats in the Strait of Hormuz. Announcing Australia's contribution of a frigate, a surveillance aircraft and station officials at the multination patrolling efforts' headquarters in Bahrain, PM Morrison described joining the effort as a matter "about freedom of shipping".[41] Following a request from the US side at the Australia–United States Ministerial meeting, nearly 200 Australians were said to be involved in the effort, with 177 personnel on the warship and 10 on the surveillance aircraft.[42]

On matters pertaining to China, despite its economic ties with the Asian giant (Australia–China trade being US$ 194.6 billion in 2017–18),[43] a shift in Australia's posture was apparent. For instance, Canberra became an early proponent of the Trump administration's global campaign against Chinese telecom giants. Taking a cue from the US and drawing on the findings of its own intelligence agencies, Australia banned Huawei Technologies Co. and ZTE Corp. from its next-generation 5G mobile network. Australia even instituted "counterespionage safeguards that China believed were directed against it, and Australian officials blocked Huawei from building an undersea cable to the Solomon Islands".[44]

Furthermore, in context of the coronavirus pandemic reportedly having originated in China, Canberra was one of the most vocal and early advocates of an independent investigation into the pandemic's origins and China's early handling of the outbreak. In addition, at the World Health Assembly for instance, Australia was one of the leading co-sponsors of a motion calling on the World Health Organization's Director General to initiate a "stepwise process of impartial, independent and comprehensive evaluation, including using existing mechanisms, as appropriate, to review experience gained and lessons learned from the WHO-coordinated international health response to Covid-19".[45] Amidst its own criticism of the WHO, the US welcomed the motion which was mostly seen as an EU-drafted but Australia-led action. Notably, US Secretary of State Mike Pompeo said that he was "heartened to see Australia, [and] other countries joining us, demanding an investigation".[46] In addition, under the Trump administration's attempt to restore the global economy and address the fallout from the coronavirus pandemic, the US president even called for an expansion of the G-7 grouping to add Australia, Russia, South Korea, and India.[47] Lastly, the Quadrilateral Security Dialogue between Japan, India, Australia, and the US witnessed a belated revival, after nearly a decade. With enhanced bilateral ties with India, Australia even joined the trilateral Malabar naval exercise between the United States, Japan, and India, to effectively lend operational heft to the Quad.[48]

Relationships teetering on the edge

On the one hand, as Japan weathered the 'Trump storm' to oversee not only the sustainment of the US–Japan alliance but also its furtherance under the Indo-Pacific matrix, the other crucial US alliance in Northeast Asia witnessed a degree of turbulence. With **South Korea** (ROK), the Trump administration had a contentious relationship over its contributions towards the stationing of US troops in ROK. Although these frictions were certainly analogous to the Trump administration's apprehensions over US troops in Japan, the stalemate over North Korea's (DPRK) denuclearisation only accentuated Trump's impulses on urgently demanding a renewed cost-sharing arrangement with Seoul.

In repeatedly threatening to withdraw US forces from South Korea, Trump demanded exorbitant rises to Seoul's annual contribution towards the troops stationed for the mutual US–ROK interest of deterring a North Korean offensive or initiation of a nuclear attack. In 2018, for instance, after continued prodding from the Trump administration, South Korea agreed to pay about US$ 1 billion, up from its US$ 800 million contribution.[49] Following this, in 2019, Trump raised the stakes to demand a staggering US$

5 billion, which the South Koreans refused to comply with.[50] Instead, they offered to increase their contribution to US$ 1.3 billion. As per revelations by Trump's former National Security Advisor John Bolton, Secretary of State Mike Pompeo and Secretary of Defense Mark Esper were both receptive of that counteroffer. However, Trump refused it, even with the US–ROK cost-sharing agreement nearing expiration in December 2019.[51] As a result, the South Koreans reportedly coughed up an additional US$ 200 million to cover the salaries of local South Koreans working on the American bases.[52] The alliance which was forged 70 years ago, seemed to teeter further as Trump's threat to withdraw forces gathered steam with his 2020 decision to slash US troop presence in Germany by a third of its strength.[53] In addition, Trump's attempt to renegotiate the predecessor administration's Korea–US Free Trade Agreement (or KORUS) only compounded frictions over burden-sharing arrangements. In exchange for voluntary export restraint on South Korean steel and an increase in annual export quotas for US auto manufacturers, the path to a KORUS 2.0 was often riddled with concerns over the Trump administration interlinking trade negotiations with its compliance to security commitments towards ROK.

Apart from these fissures, US–ROK alliance also took a hit from the standpoint of force readiness and interoperability, mainly owing to the Trump administration's focus on the issue of North Korea's nuclear programme. In contrast to the focus on DPRK playing a constructive role in US–Japan ties, US–ROK ties instead devolved. In late 2016, US President Barack Obama had warned then-President elect Donald Trump about North Korea being a "top national security priority" and an impending challenge.[54] In departing from the Obama administration's dependence on "strategic patience" and 'Left of Launch' cyber attacks,[55] the Trump administration sought to devise its approach on the North Korean issue by pursuing rigorous engagement with Pyongyang. This bore ramifications for the US–ROK dynamic. In 2018, for instance, without consulting Seoul or even the Pentagon, Trump gave in to North Korean leader Kim Jong-un's demand of overseeing a reduction in US–ROK military exercises. Following Trump's summit with the elusive leader in Singapore, the US announced the cancellation of the "large-scale, long-planned" Ulchi Freedom Guardian exercise – which the previous year "ran for 11 days and involved 17,500 American forces, including about 3,000 from outside the peninsula, and 50,000 South Korean troops".[56] The next year, following the Trump–Kim meeting in Hanoi, the US announced the suspension of two more exercises: Key Resolve and Foal Eagle. Halting the latter was particularly significant, as it traditionally includes an all-encompassing exercise of "thousands of ground, naval, air and special operations troops from both nations".[57] After having characterised such exercises as "very provocative",[58] Trump even deemed them to be "very expensive".[59]

The optics of this unravelling of the US–ROK alliance not only worsened the credibility question around the US' commitment towards allied and partner nations, it also put South Korean conservatives in a fix. Long known to be the staunchest supporters of strong ties with the US, they were then forced to re-examine their advocacy for the US–ROK alliance, especially in view of the possibility that Trump might undercut their interests to seal a politically expedient deal with DPRK – which could fall short of "complete, verifiable, and irreversible denuclearization", to simply extricate the US from its alliance commitments.[60]

Most notably, a prominent conservative South Korean lawmaker even stated, "If the United States accepts North Korea as the de facto nuclear state through a small deal, acquiring nuclear weapons is the only viable path for us".[61] North Korea as well took advantage of the situation to drive a deeper wedge between ROK and the US. Case in point is the 2020 bombing of the inter-Korean liaison office at the Kaesong Industrial Complex (KIC), which has been a symbol of North–South reconciliation.[62] By urging the South to permit the reopening of the KIC – deemed to be crucial to DPRK's export revenue – the US' attempt to fully muzzle the DPRK's economic avenues could become a bone of contention between Seoul and Washington.

Furthermore, as Bolton revealed in his book, this also forced South Korean leaders to compete with other regional leaders (chiefly, Japan's Abe) for developing a close relationship with the US president. This also worsened Japan–ROK ties either with respect to long-standing apprehensions – like the politically charged issue of Japanese reparations for South Korean 'comfort women' during the Second World War – or with nascent divergences like the simmering trade war between the two nations following South Korea's brief cessation of a military intelligence-sharing pact with Japan.[63] Such tensions only worsened the prospects of America's long-standing aim to cultivate an independent relationship between its two crucial allies in Northeast Asia. Moreover, China sensed an opportunity to stage a diplomatic coup, by urging Tokyo and Seoul to work towards finalising a trilateral free trade deal with Beijing.

Another such alliance which witnessed turbulence under Trump and posed momentary gains for China, was the US' partnership with the **Philippines**. In early 2020, tensions crescendoed with Philippine President Rodrigo Duterte announcing the termination of the Visiting Forces Agreement (VFA) with the United States. As per the 1988 agreement, the US had maintained a forward deployment in the country, conducted military exercises with local forces, and also trained and assisted the Philippines' military in counterterrorism operations against ISIS-affiliated Abu Sayyaf in Mindanao and the Sulu Sea.[64]

The decision to terminate the agreement followed heightened US criticism of Duterte's nationwide "war on drugs", which by January 2020 had left over 27,000 dead as per Human Rights Watch. Furthermore, gross human rights violations had come to light, wherein forces "routinely shot and killed alleged drug dealers and users, claiming without proof they resisted arrest" and "numerous instances in which the police planted weapons and drugs on victims to cover up summary executions".[65] The VFA termination came after the Trump administration – mostly prompted by the US Senate resolution seeking the imposition of Magnitsky Act sanctions against Duterte's officials involved in the detention of opposition Senator Leila de Lima – revoked the US visa held by Ronald Dela Rosa, a former police chief who oversaw the war.[66]

Although Trump at first welcomed Duterte's decision to terminate the VFA by saying, "[T]hank you very much, we [will] save a lot of money",[67] its termination was deemed to be a setback for bilateral ties, not only by the American security establishment but incidentally also by the Philippines' military.[68] Politically, the move came to be seen as the Philippines' increasing tilt towards China, especially after Duterte's numerous public statements against the US and his soft-peddling of Chinese actions in the South China Sea. The latter was most apparent in 2019, when Duterte announced that he would not adhere to an international court ruling (which in fact affirmed the Philippines' claims in the South China Sea) and instead would advance a joint oil exploration project with Beijing.[69]

However, in early June 2020, in possibly recognising the emergent global consensus against China following Beijing's unhindered pursuit of its strategic interest (such as announcing administrative jurisdictions over contested islands in the South China Sea) even as the world grappled with the coronavirus pandemic, the Duterte government revisited its decision. In a diplomatic letter to the US, the government explained its decision to revive the VFA "in light of political and other developments in the region".[70]

Abhorrence towards multilateralism

Beyond the United States' bilateral relationships in the region, the Trump administration's approach to the Indo-Pacific was laden with self-inflicted challenges. There was a crisis of credibility owing to the US' confrontational approach towards allied and partner nations on exacting renewed trading arrangements, its realpolitik "exclusionary policies" towards adversarial nations, and its ambivalence towards global governance challenges like climate change.[71] Most importantly, questions over the US' commitment to the Indo-Pacific strategy stemmed from the Trump administration's abhorrence for multilateralism under its 'America First' worldview. From the Paris Climate Accords to the International Criminal Court, Trump often construed the US' opposition to multilateral fora as his commitment to "never surrender America's sovereignty to an unelected, unaccountable, global bureaucracy".[72]

Moreover, three days after being sworn into office, Trump withdrew the United States from the **Trans-Pacific Partnership** (TPP), deeming the withdrawal to be a "great thing for the American worker".[73] The 12-nation free trade agreement negotiated by the Obama administration included Australia,

Brunei, Chile, New Zealand, Peru, Singapore, Vietnam, Japan, Malaysia, Canada, Mexico, and the US. Under the predecessor administration, the TPP was deemed to be a crucial component of the United States' Rebalance to the region. Notably, President Obama often argued for it in existential terms: "if we don't pass this agreement – if America doesn't write those rules – then countries like China will".[74] His Secretary of Defence Ashton Carter even likened it to the United States launching a new aircraft carrier.[75] It even received some bipartisan support in the otherwise highly partisan years of the Obama presidency. Prominent Republicans like Sens. John McCain and Marco Rubio, and Govs. John Kasich of Ohio and Jeb Bush of Florida supported the deal's advantages over increasing the US' influence in the Pacific.[76]

During the election, however, the TPP became a political lightening rod for Trump – who called it a "disaster"[77] – to berate the Washington establishment's bipartisan support for free trade. Trump had rallied against liberal internationalists in Washington for gutting America's industrial base and shipping jobs from the American Midwest to the Asia-Pacific region. Further, it didn't help that Trump's opponent and formerly Obama's Secretary of State Hillary Clinton deemed the TPP to be the "gold standard" of trade agreements. This also spurred the progressives in the Democratic Party to rally against it, irrespective of the high standards the TPP sought to set in terms of labour and environmental protections, in order to level the playing the field between the participating nations. After Trump's withdrawal, however, in 2018, the other 11 nations went on to finalise the Comprehensive and Progressive Agreement for Trans-Pacific Partnership (CPTPP) without the US.[78]

Apart from missing an opportunity to get ahead of the curve on China's inroads into the regional economy, Trump's withdrawal from the TPP also turned out to be detrimental to his campaign promise of facilitating the growth of the American manufacturing industry.

For instance, on the campaign trail, Trump tweeted: "The Trans-Pacific Partnership is an attack on America's business. It does not stop Japan's currency manipulation. This is a bad deal".[79] However, with the signing of the CPTPP without the US, American farmers stood at a greater disadvantage. What would have otherwise been the largest regional free trade agreement, the US' exclusion from the CPTPP rendered America's agricultural exports to continue to have duties imposed on them, at least until the US signed a limited deal with Japan, for instance. Hence Trump faced domestic opposition for his decision to withdraw the US from the TPP. For instance, in a joint letter to the administration's US Trade Representative Robert Lighthizer, farm groups and agro companies wrote: "U.S. exporters of wheat, beef, pork, dairy, wine, potatoes, fruits and vegetables and other products are facing collapse of their Japanese market share as these lucrative sales are handed over to their competitors".[80]

Moreover, the Trump administration's abhorrence for multilateralism also encompassed policy flubs towards regional multilateral groups, which only paved the way for an increase in Chinese influence. For instance, Trump had an ambivalent relationship with the region's prime grouping of Southeast Asian nations, i.e. the **Association of Southeast Asian Nations (ASEAN)**.

In 2019, the annual East Asian Summit and the US–ASEAN Summit witnessed a downgraded US delegation.[81] Led by Commerce Secretary Wilbur Ross and Assistant to the President for National Security Affairs Robert O'Brien, the delegation was grossly outranked by head-of-state level representation from 17 regional countries. This raised questions about the Trump administration's commitment to the Indo-Pacific region, which it had deemed as "the single most consequential region for America's future".[82] The same also signified a pattern of year-on-year downgraded representation. For example, Trump attended the 2017 US–ASEAN summit in Manila, following which Vice President Mike Pence represented the United States at its 2018 iteration in Singapore. In contrast to Obama's push for US ties with Southeast Asia, the Trump years witnessed little accomplishment on further institutionalising the US–ASEAN dynamic, for instance. However, with respect to the Trump administration's push for increased material and policy-level burden-sharing by regional partners, its offhanded approach with ASEAN encouraged that prime multilateral regional cohort to become increasingly vocal about its own positions. Often criticised for hedging between Chinese and American visions and attributing the same to its systemic challenge of coalescing ten nations around coherent policy views, ASEAN in 2019 released the "ASEAN Outlook on the Indo-Pacific". Released after more than a year of negotiations, the released document notes ASEAN's centrality in the Indo-Pacific – in an evident instance of adopting the Western nomenclature for the region – and outlines four areas of cooperation (maritime cooperation, connectivity, UN Sustainable Development Goals 2030, and economic development) to mostly echo the Trump administration's push for a region anchored in a rules-based order.[83]

Apart from this indirect – albeit significant outcome – for over three years, Trump ignored the region where "showing up" is half the battle.[84] Case in point: the Trump administration left the post of the US ambassador to ASEAN vacant. Further compounding the Trump administration's lacklustre record on this aspect was the coronavirus pandemic, spurring the postponement of the March 14, 2020, Las Vegas summit between ASEAN leaders and Trump – which would have been the first such meet on US soil under the Trump administration.

Possibly as a result of such a record, the 2020 survey of ASEAN thinkers and opinion leaders by Singapore's ISEAS–Yusof Ishak Institute had some troubling results.

In that survey, about half of the respondents (47 percent) expressed "little or no confidence in the US as a strategic partner and provider of regional security",[85] whereas a staggering 77 percent noted that US engagement with Southeast Asia "declined under the Trump administration compared to the

Obama administration".[86] However, with ASEAN releasing its outlook towards the region which evidently had a bent towards the US' aims, the survey found that if forced to pick a side, ASEAN would align with the US over China – albeit with a close 53.6 versus 46.4 percent.[87]

But when the responses were broken down by nationality, the majority of respondents – 7 of 10 ASEAN members (Brunei, Cambodia, Indonesia, Laos, Malaysia, Myanmar, and Thailand), 70 percent favoured China over the US.[88] Possibly as a result, ASEAN also staked out positions that reflected its pragmatism towards China. Prominent cases in point are the progress made on the ASEAN–China Code of Conduct (COC) in the South China and the late-2019 finalisation of the China-led Regional Comprehensive Economic Partnership (RCEP). Signed in late-2020, the latter is the "world's largest trade agreement" among the ten member nations of ASEAN and China, Japan, Australia, New Zealand, and South Korea.[89] On the COC, although it took over 15 years since the 2002 Declaration on the Conduct of Parties in the South China Sea for the two sides to declare a three-year timeframe to conclude talks,[90] the uptick in ASEAN's attention to the COC signifies its intent to prepare itself against American unreliability. Most notably, under the COC, Beijing has demanded the restriction of joint military exercises with external powers. If this demand is honoured, it would be a major blow to the US' naval projection in Southeast Asia.

Despite these setbacks to the US' influence in the region, however, the Trump administration did not shy from presenting its very own agenda for the region in order to expand America's economic influence and counter China's Belt and Road Initiative (BRI).

The American proposition

In formulating the US' response to rising Chinese influence in the Indo-Pacific region, the Trump administration reflected considerable flexibility on its 'America First' impulses. Ahead of the 2016 elections, then-candidate Trump often rallied against the US' funding of nation-building projects abroad. Most notably, he derided American aid, which in the previous years had topped the international donors list. For instance, in 2014, the US was "the biggest foreign aid donor in the world, representing about 24 percent of development assistance from major governments".[91] Furthermore, making it a prominent aspect of his campaign, in his June 2015 speech (in which he announced his candidacy), Trump rallied against foreign aid and the US' "rebuilding" of other nations, and pledged to invest in US infrastructure instead.[92]

However, upon taking office, Trump came around to encouraging development finance and regional infrastructure projects with partner nations.

Under its Belt and Road Initiative, Beijing has offered development assistance for regional infrastructure. With its intended investment of US$ 1 trillion, China has focused on developing ports, roads, railways, airports, power plants, and even telecommunication networks.[93] However, in face of allegations of employing its investments to 'debt-trap' recipient nations for strategic purposes, China has sought to dampen the stigma around its lending. In doing so, it has repurposed the BRI's commitment as being towards "open, green and clean" investments. However, according to the discussed ISEAS–Yusof Ishak Institute survey, most respondents in Southeast Asia (63.6 percent) have "little or no confidence in the new BRI approach".[94]

Hence, in building on the normative focus of the Indo-Pacific strategy, the Trump administration sought to counter China's investment programmes. The US announced a slew of initiatives to further liberal market values, with their focus on anti-corruption measures, standards for fiscal transparency, offering assistance towards youth development and media freedom, and protecting fundamental freedoms and human rights.[95] The Trump administration also did not shy from openly contrasting its approach vis-à-vis China. Notably, at the APEC summit in 2018, Vice President Mike Pence said, "We don't drown our partners in a sea of debt. We don't coerce, corrupt or compromise your independence".[96]

Charting an alternate economic model for the region

One prominent instance of the Trump administration dampening its nativist agenda came in late 2018. Following a bipartisan effort in the US Congress, the US president signed into law the Better Utilization of Investments Leading to Development Act of 2018 (BUILD Act).[97] The Act mandated the executive branch to establish the US International Development Finance Corporation (DFC) by "consolidating and expanding existing U.S. government development finance functions",[98] which until then were conducted chiefly by the Overseas Private Investment Corporation (OPIC) and other departments of the US Agency for International Development (USAID). In subsuming the OPIC's duties – which was established back in 1971 as "a lending facility to encourage American companies to invest in developing countries",[99] the DFC was accorded an enhanced budget of US$ 60 billion (twice the erstwhile department's lending capacity of about US$ 29 billion).

In advancing the US' influence in the region by "incentivizing private investment as an alternative to a state-directed investment model", the Act mandated the DFC to have more "tools" to provide investment support.[100] Relative to the OPIC, the DFC has the authority to offer limited equity investments with due technical assistance; operates with a longer authorisation period, i.e. up to seven years vis-à-vis OPIC's year-to-year authorisation; and has oversight and risk management mechanisms like having its own inspector general (IG), as compared to OPIC (which fell under the USAID IG's jurisdiction).

Interestingly, just a year before Trump signed the BUILD Act into law, his administration had proposed gutting the US State Department's budget by one-third, wherein OPIC's budget was completely slashed. Although the US Congress later rejected those proposed cuts, Capitol Hill itself took its time to come around to the idea of combating Chinese influence with development aid. Case in point, a few years earlier, a gridlocked Congress had little interest in President Obama's effort to establish an American development finance bank inline with his administration's elevation of development assistance as a core pillar of US foreign policy alongside hard power tools.[101] Thus, the legislation passed under Trump invited surprise from many US Congressmen as well. For instance, after the BUILD Act was signed into law, Sen. Chris Coons (D-DE) said, "I'm astonished, to be honest. I still can't believe we got it done. . . . We've been working on this since 2015. It is basically the same proposal we had under the Obama administration. We rebranded to focus it on China".[102]

This reinvigorated effort to push American development aid by seeing it through with the rationale of confronting China underscored the emergent US bipartisan consensus on China and the Trump administration's role in ending the post–Cold War dissonance on the US' China policy.

Another pivotal move facilitated by the US Congress came in mid-2019, when legislators confirmed the board for the Export-Import Bank of the United States (EXIM). The same restored programmes worth US$ 40 billion of potential US export deals estimated to support nearly 230,000 US jobs – thereby also bearing gains for Trump's 2016 pledge to spur growth in American manufacturing jobs. The same occurred after prodding from Trump's Director of the National Economic Council Larry Kudlow, who termed EXIM as "a financial tool and a national security weapon" against China.[103] Awaiting action since July 2015, the US Senate accorded EXIM the quorum required "to approve financing or loan guarantees larger than $10 million".[104] As a result, the Trump administration eventually reported US exports (goods and services) to the Indo-Pacific to have breached US$ 720 billion, up by 4.5 percent over 2017 figures.[105]

Moreover, the Trump administration also included the showcasing of American investments' high-standards as part of the "economic pillar" of the Indo-Pacific strategy.[106] For instance, together with traditional partners like Japan and Australia, the US also announced the Blue Dot Network (BDN) in its effort "to certify high-quality infrastructure". Wherein, the Trump administration sought to provide a "globally recognized seal of approval signifying adherence to high standards", and to "promote market-driven, transparent, and financially sustainable infrastructure projects".[107] This also translated into the Trump administration dampening its abhorrence for multilateral platforms, as the BDN encompassed a set of cooperative arrangements put into force in 2017 between DFC and its counterpart development finance institutions in Japan, Australia, Singapore, Canada, and the European Union. Furthermore, with Infrastructure Finance and Market Building Agreements with South Korea and Singapore, the Trump administration sought to attract "private sector financing, develop local debt and capital markets, bolster utility creditworthiness, and attract institutional investors" into the region's infrastructure projects.[108]

Furthermore, the Trump administration also encouraged US investments in the Indo-Pacific region, for the mutual benefit of the region and US businesses looking to capitalise on the 'Asian century'. Under the Indo-Pacific Business Forum (IPBF) launched in mid-2018, for instance, the Trump administration committed US$ 113 million to seed the region's digital economy, energy, and infrastructure projects.[109] Under this initiative, the US Department of Commerce also "assisted more than 9,000 U.S. companies to do business in the Indo-Pacific region and facilitated more than $7.65 billion in U.S. exports, of which, $3.1 billion was for government tenders won by U.S. firms".[110] Similarly, it also "assisted over 2,500 Indo-Pacific companies seeking to invest in the United States facilitating $18 billion of inward investment into the United States from the Indo-Pacific".[111]

Moreover, under its policy of "unleashing American energy dominance" via tapping into "new export opportunities" for American energy producers,[112] the Trump administration turned to the Indo-Pacific – which is expected to account for a staggering 60 percent of global growth in energy demand by 2040.[113] With about a third of US energy exports worth US$ 50 billion already reaching the region,[114] the administration announced the Asia Enhancing Development and Growth through Energy (Asia EDGE) initiative to "grow sustainable and secure energy markets throughout the Indo-Pacific".[115] With Asia EDGE, the US speared the development of energy infrastructure in the region. For instance, the initiative "mobilized $806 million in public and private investment across 11 renewable energy projects" in Indonesia alone.[116] With the Asian Development Bank, USAID aims to mobilise up to US$ 7 billion in investment to reform the region's energy sector.[117]

As with DFC, the US under this initiative has also encouraged partnerships with nations like Japan – which has pledged US$ 10 billion towards regional LNG development.[118] For instance, with their co-financing agreement, US EXIM and Japan's Nippon Export and Investment Insurance (NEXI) have facilitated financing of liquified natural gas (LNG) shipments to the region.[119] Another prominent example is the Trilateral Infrastructure Partnership between Japan, Australia, and the US, which includes projects worth US$ 400 million to support the Papua New Guinea Electrification Partnership.[120]

Finally, beyond energy infrastructure, under the Trump administration's Infrastructure Transaction and Assistance Network (ITAN), the US also assumed an advisory role in catering to regional nations' need for expertise. For instance, Washington has extended support on strengthening the legal and regulatory framework for infrastructure in Bangladesh and Nepal, assisted the Philippines with its infrastructure development strategy, and advised Vietnam on its announced Power Development Plan.[121]

Beyond this shift in the 'America First' worldview over development support and finance, the Trump administration also kept its abhorrence for security commitments at bay with respect to cultivating nascent partners in the region.

Cultivating nascent partners

Owing to the predecessor administration's rapprochement with Vietnam – primarily with the US lifting its decades-long arms embargo and facilitating its entry into the TPP – Hanoi gained strategic currency with Washington. Beyond rising China–Vietnam tensions with respect to their claims in the South China Sea, US–Vietnam ties have also benefited at the hands of favourable opinion amongst the Vietnamese. According to a 2015 Pew Research Center survey, "76 percent of Vietnamese had 'favorable' views of the US, which was even higher [89 percent] among 'more highly educated people.' It was one of the highest such percentages of any country included in the poll".[122] Hence, under Trump, the country was reported as "Washington's new darling".[123] Since assuming office, Trump visited the country twice: first, for the discussed APEC summit in November 2017, where the US president first presented the US' vision for the Indo-Pacific region; second, for his follow-up meeting with North Korean leader Kim Jong-un in early 2019. Trump also hosted Vietnamese Prime Minister Nguyen Xuan Phuc in early 2017, as the first Southeast Asian leader invited to the White House under the Trump administration.

Moreover, Washington accorded a strategic dimension to US–Vietnam ties, with the March 2018 visit of the USS *Carl Vinson* to Vietnam – the first US aircraft carrier to have conducted such a visit since the end of the Vietnam War in 1975 and normalisation of diplomatic ties in 1995. Thereafter, the US completed its second aircraft carrier visit in March 2020.[124] Furthermore, the Trump administration oversaw the transfer of patrol boats and equipment worth over US$ 12 million to the Vietnamese Coast Guard[125] and sealed an overall deal to provide 24 forty-five-foot Metal Shark patrol boats.[126]

Under Trump, Vietnam also assumed geoeconomic importance, with American firms increasingly viewing the Southeast Asian nation as an ideal destination to move operations out of China. Over the course of the 18-month US–China trade war, gains for Vietnam have been evident in terms of unprecedented rise in its bilateral trade with the US. America's imports from Vietnam for instance, surged by "36 percent, or $17.5 billion".[127] In broader terms, this proved instrumental in raising bilateral trade through the US$ 77 billion mark last year – up from just a little over US$ 400 million in 1994.[128]

This also spurred a levelling-up of US commitments towards Hanoi. For instance, USAID "supported Vietnam's reform of over 180 commercial laws and regulations", which it now attributes to Vietnam's competitiveness ranking in the World Bank's "Doing Business" report rising from 82 in 2017 to 68 in 2018.[129] Furthermore, in 2020, the two sides signed a US$ 42 million agreement on assisting Vietnam's Ministry of Planning and Investment (MPI) to enhance its "economic competitiveness through strengthening the competitiveness of the private sector and promoting innovation, start-up ecosystems and human capital".[130]

The Trump administration also focused on Indonesia. In recognising the traditional dependence of Indonesia, Vietnam and India on Russian weaponry, the Trump administration oversaw the passage of Section 231 waiver provisions under the Countering America's Adversaries Through Sanctions Act (CAATSA) in order to accord the three countries some leeway on their slow – yet substantial, progress on reducing the share of Russian arms in their respective arsenals.[131] Moreover, Indonesia was also identified as a geoeconomic partner much like Vietnam. Following a phone call between Indonesian President Joko Widodo and his US counterpart, Central Java Governor Ganjar Pranowo announced the clearing up of "4,000 hectares of land in Central Java to accommodate US companies planning to relocate from China in the wake of the coronavirus pandemic and an escalating trade war".[132]

Similarly, with Singapore, strategic ties strengthened with the Trump administration renewing the 1990 'Memorandum of Understanding Regarding United States Use of Facilities in Singapore' – deemed as "a landmark agreement which underpinned America's security presence in the region for almost 30 years".[133] With the two sides agreeing to amend the MoU, American forces' use of Singapore's air and naval bases stands extended by another 15 years (until 2035). Under it, the US has been rotationally deploying fighter aircrafts for exercises and other purposes. Under Trump, the US State Department also approved the sale of the F-35 Joint Strike fighter jets to Singapore – making it only the 12th country to purchase the highly coveted fifth-generation advanced-stealth fighter aircraft.[134]

Beyond focussing on these Southeast Asian nations, another nascent US partner with which the Trump administration practiced considerable continuity and also elevated bilateral ties with force interoperability agreements and institutionalisation of relations was India. Before examining the US–India strategic partnership, however, a perusal of the Trump administration's approach towards the larger South Asian subregion of the Indo-Pacific stands warranted.

Notes

1 Donald Trump, "Remarks by President Trump at APEC CEO Summit", *US Embassy and Consulate in Vietnam*, 10 November 2017, last accessed 20 September 2020, https://vn.usembassy.gov/20171110-remarks-president-trump-apec-ceo-summit/

2 *Quoted in* Tara Copp, "INDOPACOM, It Is: US Pacific Command Gets Renamed", *Defense News*, 30 May 2018, last accessed 20 September 2020, www.defensenews.com/news/your-military/2018/05/30/indo-pacom-it-is-pacific-command-gets-renamed/

3 NSS 2017, "National Security Strategy of the United States of America", *The Donald Trump Administration – The White House Archives*, December 2017, pp. 45–6, last accessed 20 September 2020, www.whitehouse.gov/wp-content/uploads/2017/12/NSS-Final-12-18-2017-0905.pdf

4 State Department report, "A Free and Open Indo-Pacific – Advancing a Shared Vision", *The US Department of State*, 04 November 2019, p. 6, last accessed 20 September 2020, www.state.gov/wp-content/uploads/2019/11/Free-and-Open-Indo-Pacific-4Nov2019.pdf

5 State Department report, "A Free and Open Indo-Pacific – Advancing a Shared Vision", *The US Department of State*, 04 November 2019, p. 4, last accessed 20 September 2020, www.state.gov/wp-content/uploads/2019/11/Free-and-Open-Indo-Pacific-4Nov2019.pdf

6 State Department report, "A Free and Open Indo-Pacific – Advancing a Shared Vision", *The US Department of State*, 04 November 2019, p. 12, last accessed 20 September 2020, www.state.gov/wp-content/uploads/2019/11/Free-and-Open-Indo-Pacific-4Nov2019.pdf

7 State Department report, "A Free and Open Indo-Pacific – Advancing a Shared Vision", *The US Department of State*, 04 November 2019, p. 13, last accessed 20 September 2020, www.state.gov/wp-content/uploads/2019/11/Free-and-Open-Indo-Pacific-4Nov2019.pdf

8 State Department report, "A Free and Open Indo-Pacific – Advancing a Shared Vision", *The US Department of State*, 04 November 2019, p. 21, last accessed 20 September 2020, www.state.gov/wp-content/uploads/2019/11/Free-and-Open-Indo-Pacific-4Nov2019.pdf

9 State Department report, "A Free and Open Indo-Pacific – Advancing a Shared Vision", *The US Department of State*, 04 November 2019, p. 29, last accessed 20 September 2020, www.state.gov/wp-content/uploads/2019/11/Free-and-Open-Indo-Pacific-4Nov2019.pdf

10 DoD report, "Summary of the 2018 National Defense Strategy of the United States of America – Sharpening the American Military's Competitive Edge", *The US Department of Defense*, n.d., last accessed 20 September 2020, https://dod.defense.gov/Portals/1/Documents/pubs/2018-National-Defense-Strategy-Summary.pdf

11 DoD report, "Indo-Pacific Strategy Report – Preparedness, Partnerships, and Promoting a Networked Region", *The US Department of Defense*, 01 June 2019, last accessed 20 September 2020, https://media.defense.gov/2019/Jul/01/2002152311/-1/-1/1/DEPARTMENT-OF-DEFENSE-INDO-PACIFIC-STRATEGY-REPORT-2019.PDF

12 DoD report, "Summary of the 2018 National Defense Strategy of the United States of America – Sharpening the American Military's Competitive Edge", *The US Department of Defense*, n.d., p. 9, last accessed 20 September 2020, https://dod.defense.gov/Portals/1/Documents/pubs/2018-National-Defense-Strategy-Summary.pdf

13 David Shambaugh, "China Engages Asia: Reshaping the Regional Order", *International Security*, Volume 29, Issue 3 (Winter 2004/05), p. 95.

14 Hannah Fodale, "RESOLVED: Abe's Investment in His Relationship with President Trump Has Advanced Japanese Interests", *Center for Strategic and International Studies*, 30 January 2020, last accessed 20 September 2020, www.csis.org/analysis/resolved-abes-investment-his-relationship-president-trump-has-advanced-japanese-interests

15 Sheila Smith, "Forty-two and Counting . . . The Trump–Abe Connection", *Council on Foreign Relations*, 23 May 2019, last accessed 20 September 2020, www.cfr.org/blog/forty-two-and-countingthe-trump-abe-connection

16 Ayako Mie, "How Much Does Japan Pay to Host U.S. Forces? Depends on Who You Ask", *The Japan Times*, 31 January 2017, last accessed 20 September 2020, www.japantimes.co.jp/news/2017/01/31/national/much-japan-pay-host-u-s-forces-depends-ask/#.XhKNsRczZTY
17 Maggie Haberman and David Sanger, "Transcript: Donald Trump Expounds on His Foreign Policy Views", *The New York Times*, 26 March 2016, last accessed 20 September 2020, www.nytimes.com/2016/03/27/us/politics/donald-trump-transcript.html
18 *Quoted in* BBC report, "Japan PM Is First Foreign Leader to Meet Trump", *BBC*, 17 November 2016, last accessed 20 September 2020, www.bbc.com/news/world-asia-37946613
19 *Quoted in* Reuters report, "Trump Says U.S. Behind Japan '100 Percent' after North Korea Missile Launch", *Reuters*, 12 February 2017, last accessed 20 September 2020, www.reuters.com/article/us-northkorea-missiles-trump-idUSKBN15R05E
20 Mari Yamaguchi, "Trump Meets with Families of Japanese Abducted by N. Korea", *ABC News*, 27 May 2019, last accessed 20 September 2020, https://abcnews.go.com/International/wireStory/trump-meets-families-japanese-abducted-korea-63300164
21 Donald Trump, "Remarks by President Trump to the 72nd Session of the United Nations General Assembly", *The Donald Trump Administration – The White House Archives*, 19 September 2017, last accessed 20 September 2020, www.whitehouse.gov/briefings-statements/remarks-president-trump-72nd-session-united-nations-general-assembly/
22 Donald Trump, "Remarks by President Trump and Prime Minister Abe of Japan in Joint Press Conference", *The Donald Trump Administration – The White House Archives*, 27 May 2019, last accessed 20 September 2020, www.whitehouse.gov/briefings-statements/remarks-president-trump-prime-minister-abe-japan-joint-press-conference-3/
23 *Quoted in* Brad Lendon, "Mattis: US Will Defend Japanese Islands Claimed by China", *CNN*, 04 February 2017, last accessed 20 September 2020, https://edition.cnn.com/2017/02/03/asia/us-defense-secretary-mattis-japan-visit/index.html
24 Don Lee, "Trump's Withdrawal from TPP Trade Deal Is Hurting U.S. Exports to Japan", *LA Times*, 25 April 2019, last accessed 20 September 2020, www.latimes.com/politics/la-na-pol-trump-abe-us-japan-trade-tpp-20190425-story.html
25 Paul Wiseman, "US Signs Limited Deal with Japan on ag, Digital Trade", *Associated Press*, 08 October 2019, last accessed 20 September 2020, https://apnews.com/5c7adfefc4824bcba15daf977250fb9b
26 Donald Trump, "Remarks by President Trump at Signing of the U.S.–Japan Trade Agreement and U.S.–Japan Digital Trade Agreement", *The Donald Trump Administration – The White House Archives*, 07 October 2019, last accessed 20 September 2020, www.whitehouse.gov/briefings-statements/remarks-president-trump-signing-u-s-japan-trade-agreement-u-s-japan-digital-trade-agreement/
27 Matthew Goodman, "Scoring the Trump–Abe Trade Deal", *Center for Strategic and International Studies*, 01 October 2019, last accessed 20 September 2020, www.csis.org/analysis/scoring-trump-abe-trade-deal
28 Paul Wiseman, "US Signs Limited Deal with Japan on ag, Digital Trade", *Associated Press*, 08 October 2019, last accessed 20 September 2020, https://apnews.com/5c7adfefc4824bcba15daf977250fb9b
29 Franz-Stefan Gady, "Japan to Continue Assembling F-35A Fighter Jets at Home", *The Diplomat*, 08 January 2020, last accessed 20 September 2020, https://thediplomat.com/2020/01/japan-to-continue-assembling-f-35a-fighter-jets-at-home/
30 Aaron Mehta, "With Massive F-35 Increase, Japan Is Now Biggest International Buyer", *Defense News*, 18 December 2018, last accessed 20 September 2020,

www.defensenews.com/global/asia-pacific/2018/12/18/with-massive-f-35-increase-japan-is-now-biggest-international-buyer/

31 Aaron Mehta, Valerie Insinna, and Mike Yeo, "Japan Wants to Be an Official F-35 Partner. The Pentagon Plans to Say No.", *Defense News*, 29 July 2019, last accessed 20 September 2020, www.defensenews.com/global/asia-pacific/2019/07/29/japan-wants-to-be-an-official-f-35-partner-the-pentagon-plans-to-say-no/

32 Michael Bender, "Trump Urges Japan to Buy More U.S. Military Equipment", *The Wall Street Journal*, 07 November 2017, last accessed 20 September 2020, https://accounts.wsj.com/login?target=https%3A%2F%2Fwww.wsj.com%2Farticles%2Ftrump-presses-japan-for-bilateral-trade-deal-1509938809

33 Shinzo Abe, "Address by Prime Minister Shinzo Abe at the Opening Session of the Sixth Tokyo International Conference on African Development (TICAD VI)", *Ministry of Foreign Affairs of Japan*, 27 August 2016, last accessed 20 September 2020, www.mofa.go.jp/afr/af2/page4e_000496.html

34 Thomas Wilkins, "Australia and the Indo-Pacific: A Region in Search of a Strategy, or a Strategy in Search of a Region?", *Italian Institute for International Political Studies*, 04 June 2018, last accessed 20 September 2020, www.ispionline.it/it/pubblicazione/australia-and-indo-pacific-region-search-strategy-or-strategy-search-region-20694

35 ABC report, "Donald Trump Grants His Second State Dinner at White House to Scott Morrison", *ABC News*, 22 July 2019, last accessed 20 September 2020, www.abc.net.au/news/2019-07-23/scott-morrison-trump-state-dinner/11336950

36 Samuel Parmer, "The U.S.–Australia Alliance: What to Know", *Council on Foreign Relations*, 13 September 2019, last accessed 20 September 2020, www.cfr.org/in-brief/us-australia-alliance-what-know

37 Gina Harkins, "There Are Now 2,500 US Marines in Australia After Years of Buildup", *Military*, 25 July 2019, last accessed 20 September 2020, www.military.com/daily-news/2019/07/25/there-are-now-2500-us-marines-australia-after-years-buildup.html

38 *Quoted in* Gina Harkins, "There Are Now 2,500 US Marines in Australia After Years of Buildup", *Military*, 25 July 2019, last accessed 20 September 2020, www.military.com/daily-news/2019/07/25/there-are-now-2500-us-marines-australia-after-years-buildup.html

39 State Department factsheet, "U.S. Relations with Australia", *The US Department of State*, 21 January 2020, last accessed 20 September 2020, www.state.gov/u-s-relations-with-australia/

40 Louis Dillon, "Largest Ever Exercise Talisman Sabre Comes to an End", *Defence Connect*, 30 July 2019, last accessed 20 September 2020, www.defenceconnect.com.au/key-enablers/4496-largest-ever-exercise-talisman-sabre-comes-to-an-end

41 *Quoted in* Jamie Tarabay, "Australia Is Third Country to Join U.S. in Patrolling Strait of Hormuz", *The New York Times*, 21 August 2019, last accessed 20 September 2020, www.nytimes.com/2019/08/21/world/australia/ships-hormuz.html

42 Graeme Dobell, "Scott Morrison, Donald Trump, and the Indo-Pacific", *Comparative Connections*, September 2019, last accessed 20 September 2020, http://cc.pacforum.org/2019/09/scott-morrison-donald-trump-and-the-indo-pacific/

43 DFAT report, "Trade and Investment at a Glance", *Department of Foreign Affairs and Trade*, 2019, p. 14, last accessed 20 September 2020, www.dfat.gov.au/sites/default/files/trade-and-investment-at-a-glance-2019.pdf

44 Mike Cherney and Dan Strumpf, "Taking Cue from the U.S., Australia Bans Huawei from 5G Network", *The Wall Street Journal*, 23 August 2018, last accessed 20 September 2020, www.wsj.com/articles/australia-bans-chinas-huawei-from-5g-network-rollout-1534992631

45 Daniel Hurst, "Australia Hails Global Support for Independent Coronavirus Investigation", *The Guardian*, 18 May 2020, last accessed 20 September 2020, www.theguardian.com/world/2020/may/18/australia-wins-international-support-for-independent-coronavirus-inquiry
46 Daniel Flitton, "Australia: Is the United States Still a Reliable Ally?", *Council on Foreign Relations*, 11 May 2020, last accessed 20 September 2020, www.cfr.org/blog/australia-united-states-still-reliable-ally
47 Anthony Galloway and Rob Harris, "Australia Set to Be Part of Trump's G7 Expansion", *Sydney Morning Herald*, 31 May 2020, last accessed 20 September 2020, www.smh.com.au/politics/federal/australia-set-to-be-part-of-trump-s-g7-expansion-20200531-p54y36.html
48 Harsh Pant and Premesha Saha, "India's Pivot to Australia", *Foreign Policy*, 21 July 2020, last accessed 20 September 2020, https://foreignpolicy.com/2020/07/21/indias-pivot-to-australia/
49 Ryan Browne, "US and South Korea Fail to Reach Cost Sharing Agreement for US Troops", *CNN*, 18 December 2019, last accessed 20 September 2020, https://edition.cnn.com/2019/12/18/politics/us-south-korea-troop-cost-sharing-fail/index.html
50 Josh Smith, "Seoul, Washington Fail to Agree on Cost of U.S. Troops; U.S. 'Not Focused on' $5 Billion Demand", *Reuters*, 18 December 2019, last accessed 20 September 2020, www.reuters.com/article/us-southkorea-usa-idUSKBN1YM0WP
51 Sue Mi Terry, "The Unraveling of the U.S.–South Korean Alliance", *Foreign Affairs*, 03 July 2020, last accessed 20 September 2020, www.foreignaffairs.com/articles/north-korea/2020-07-03/unraveling-us-south-korean-alliance
52 William Gallo, "US, South Korea Agree to Fund Furloughed Workers on US Bases", *US News*, 03 June 2020, last accessed 20 September 2020, www.voanews.com/usa/us-south-korea-agree-fund-furloughed-workers-us-bases
53 Karen DeYoung, "Trump Approves Plan to Withdraw One-third of U.S. Troops from Germany", *The Washington Post*, 06 June 2020, last accessed 20 September 2020, www.washingtonpost.com/national-security/trump-approves-plan-to-withdraw-one-third-of-us-troops-from-germany/2020/06/05/987b2d38-a775–11ea-b473–04905b1af82b_story.html
54 Christine Wang, "Barack Obama Warned Donald Trump About One Very Specific Problem and It Looks Like It's North Korea", *CNBC*, 23 November 2016, last accessed 20 September 2020, www.cnbc.com/2016/11/23/barack-obama-warned-donald-trump-about-one-very-specific-problem-and-it-looks-like-its-north-korea.html
55 David Sanger and William Broad, "Trump Inherits a Secret Cyberwar Against North Korean Missiles", *The New York Times*, 04 March 2017, last accessed 20 September 2020, www.nytimes.com/2017/03/04/world/asia/north-korea-missile-program-sabotage.html
56 Eric Schmitt, "Pentagon Suspends Major War Game with South Korea", *The New York Times*, 18 June 2018, last accessed 20 September 2020, www.nytimes.com/2018/06/18/us/politics/pentagon-south-korea-war-game-suspended.html
57 Helene Cooper, "Pentagon Again Suspends Large-scale Military Exercises with South Korea", *The New York Times*, 01 March 2019, last accessed 20 September 2020, www.nytimes.com/2019/03/01/world/asia/us-military-exercises-south-korea.html
58 *Quoted in* Eric Schmitt, "Pentagon Suspends Major War Game with South Korea", *The New York Times*, 18 June 2018, last accessed 20 September 2020, www.nytimes.com/2018/06/18/us/politics/pentagon-south-korea-war-game-suspended.html

59 *Quoted in* Helene Cooper, "Pentagon Again Suspends Large-Scale Military Exercises with South Korea", *The New York Times*, 01 March 2019, last accessed 20 September 2020, www.nytimes.com/2019/03/01/world/asia/us-military-exercises-south-korea.html
60 Eun A Jo, "The Remaking of South Korean Conservatism in the Age of Trump", *National Interest*, 30 October 2019, last accessed 20 September 2020, https://nationalinterest.org/blog/korea-watch/remaking-south-korean-conservatism-age-trump-92256
61 *Quoted in* Youkyung Lee, "South Korea Opposition Leader Sees Danger in Weak Nukes Deal", *Bloomberg Quint*, 02 February 2019, last accessed 20 September 2020, www.bloombergquint.com/business/south-korea-opposition-leader-sees-danger-in-weak-nuclear-deal
62 Choe Sang-Hun, "North Korea's Wrecking of Liaison Office a 'Death Knell' for Ties with the South", *The New York Times*, 16 June 2020, last accessed 20 September 2020, www.nytimes.com/2020/06/16/world/asia/north-korea-explosion-liaison-office.html
63 BBC report, "South Korea to Continue Intelligence-sharing Pact with Japan", *BBC*, 23 November 2019, last accessed 20 September 2020, www.bbc.com/news/world-asia-50526883
64 Nick Aspinwall, "Duterte Terminates U.S. Defense Pact, Pleasing Trump but Few Others", *Foreign Policy*, 14 February 2020, last accessed 20 September 2020, https://foreignpolicy.com/2020/02/14/vfa-philippines-china-duterte-terminates-us-defense-pact-trump/
65 John Sifton, "US Revokes Visa of Philippines 'Drug War' Police Chief", *Human Rights Watch*, 22 January 2020, last accessed 20 September 2020, www.hrw.org/news/2020/01/22/us-revokes-visa-philippines-drug-war-police-chief
66 Eimor Santos, "US Senate Asks Trump to Sanction Filipino Officials Linked to Killings, De Lima Detention", *CNN*, 09 January 2020, last accessed 20 September 2020, https://cnnphilippines.com/news/2020/1/9/united-states-senate-sanctions-philippine-officials-de-lima-drug-war.html
67 Patsy Widakuswara, "Trump Downplays Philippines' Termination of Military Pact", *VOA News*, 12 February 2020, last accessed 20 September 2020, www.voanews.com/usa/trump-downplays-philippines-termination-military-pact
68 JC Gotinga, "Military Members Disagree with VFA Termination – Senior Officers", *Rappler*, 11 February 2020, last accessed 20 September 2020, https://rappler.com/newsbreak/inside-track/senior-officers-say-military-members-disagree-vfa-termination
69 Andreo Calonzo, "Duterte Will Ignore South China Sea Ruling for China Oil Deal", *Bloomberg*, 11 September 2019, last accessed 20 September 2020, www.bloomberg.com/news/articles/2019-09-11/duterte-will-ignore-south-china-sea-ruling-for-china-oil-deal
70 VOA report, "Philippines Reinstates Pact with US Military", *VOA News*, 03 June 2020, last accessed 20 September 2020, www.voanews.com/east-asia-pacific/philippines-reinstates-pact-us-military
71 Elliot Silverberg and Matthew Sullivan, "Assessing Trump's Indo-Pacific Strategy, 2 Years In", *The Diplomat*, 01 October 2019, last accessed 20 September 2020, https://thediplomat.com/2019/10/assessing-trumps-indo-pacific-strategy-2-years-in/
72 Donald Trump, "Remarks by President Trump to the 73rd Session of the United Nations General Assembly | New York, NY", *The Donald Trump Administration – The White House Archives*, 25 September 2018, last accessed 20 September 2020,

www.whitehouse.gov/briefings-statements/remarks-president-trump-73rd-session-united-nations-general-assembly-new-york-ny/

73 *Quoted in* Nicholas Loffredo, "Trump Says Withdrawal from TPP a 'Great Thing for the American Worker'", *Newsweek*, 23 January 2017, last accessed 20 September 2020, www.newsweek.com/trump-withdrawal-tpp-great-thing-american-worker-547109

74 *Quoted in* David Nakamura and Mike DeBonis, "The Battle over President Obama's Trade Deal Has Officially Arrived", *The Washington Post*, 05 November 2015, last accessed 20 September 2020, www.washingtonpost.com/politics/obama-administration-prepares-to-launch-long-fight-over-trade-pact/2015/11/05/8299ad32-8326–11e5–8ba6-cec48b74b2a7_story.html

75 Prashanth Parameswaran, "TPP as Important as Another Aircraft Carrier: US Defense Secretary", *The Diplomat*, 08 April 2015, last accessed 20 September 2020, https://thediplomat.com/2015/04/tpp-as-important-as-another-aircraft-carrier-us-defense-secretary/

76 Everett Rosenfeld, "Trans-Pacific Partnership: GOP Candidates Split on Deal", *CNBC*, 21 October 2015, last accessed 20 September 2020, www.cnbc.com/2015/10/21/trans-pacific-partnership-gop-candidates-split-on-deal.html

77 *Quoted in* Leslie Shaffer and Everett Rosenfeld, "Trump Calls It a 'Disaster,' but Top Experts Are Saying TPP Might Still Happen Without Him", *CNBC*, 08 May 2017, last accessed 20 September 2020, www.cnbc.com/2017/05/08/trump-calls-it-a-disaster-but-top-experts-are-saying-tpp-might-still-happen-without-him.html

78 Kaite Lobosco, "Trump Pulled out of a Massive Trade Deal. Now 11 Countries Are Going Ahead Without the US", *CNN*, 30 December 2018, last accessed 20 September 2020, https://edition.cnn.com/2018/12/29/politics/tpp-trade-trump/index.html

79 Donald Trump, "The Trans-Pacific Partnership Is an Attack on America's Business. It Does Not Stop Japan's Currency Manipulation. This Is a Bad Deal", *Donald Trump – Twitter*, 23 April 2015, last accessed 20 September 2020, https://twitter.com/realDonaldTrump/status/590982593736286208?s=20

80 Don Lett, "Trump's Withdrawal from TPP Trade Deal Is Hurting U.S. Exports to Japan", *LA Times*, 25 April 2019, last accessed 20 September 2020, www.latimes.com/politics/la-na-pol-trump-abe-us-japan-trade-tpp-20190425-story.html

81 David Brunnstrom, "U.S. Set to Disappoint Asia with Downgraded Delegation for Bangkok Summits", *Reuters*, 30 October 2019, last accessed 20 September 2020, www.reuters.com/article/us-usa-asia/us-set-to-disappoint-asia-with-downgraded-delegation-for-bangkok-summits-idUSKBN1X9023

82 DoD report, "Indo-Pacific Strategy Report – Preparedness, Partnerships, and Promoting a Networked Region", *The US Department of Defense*, 01 June 2019, p. 1, last accessed 20 September 2020, https://media.defense.gov/2019/Jul/01/2002152311/-1/-1/1/DEPARTMENT-OF-DEFENSE-INDO-PACIFIC-STRATEGY-REPORT-2019.PDF

83 ASEAN report, "ASEAN Outlook on the Indo-Pacific", *ASEAN*, 23 June 2019, p. 2, last accessed 21 September 2020, https://asean.org/storage/2019/06/ASEAN-Outlook-on-the-Indo-Pacific_FINAL_22062019.pdf

84 Charles Dunst, "How the US Can Win Back Southeast Asia", *The Diplomat*, 01 September 2020, last accessed 21 September 2020, https://thediplomat.com/2020/09/how-the-us-can-win-back-southeast-asia/

85 ISEAS–Yusof Ishak Institute report, "The State of Southeast Asia 2020", *ASEAN Studies Centre–Yusof Ishak Institute*, 16 January 2020, p. 2, last accessed 21

September 2020, www.iseas.edu.sg/wp-content/uploads/pdfs/TheStateofSEA-SurveyReport_2020.pdf

86 ISEAS–Yusof Ishak Institute report, "The State of Southeast Asia 2020", *ASEAN Studies Centre–Yusof Ishak Institute*, 16 January 2020, p. 2, last accessed 21 September 2020, www.iseas.edu.sg/wp-content/uploads/pdfs/TheStateofSEA-SurveyReport_2020.pdf

87 ISEAS–Yusof Ishak Institute report, "The State of Southeast Asia 2020", *ASEAN Studies Centre – Yusof Ishak Institute*, 16 January 2020, p. 29, last accessed 21 September 2020, www.iseas.edu.sg/wp-content/uploads/pdfs/TheStateofSEASurveyReport_2020.pdf

88 Joshua Kurlantzick, "U.S.–ASEAN Relations – No Summit, But What's the Status", *Council on Foreign Relations*, 02 March 2020, last accessed 20 September 2020, www.cfr.org/blog/us-asean-relations-no-summit-whats-status

89 ASEAN press release, "RCEP Leaders Agree to Sign Trade Pact in 2020", *ASEAN*, 07 November 2019, last accessed 20 September 2020, https://asean.org/rcep-leaders-agree-sign-trade-pact-2020/

90 Tommy Chai, "ASEAN–China Code of Conduct to Conclude in Three Years?", *Foreign Brief*, 16 November 2018, last accessed 20 September 2020, www.foreignbrief.com/asia-pacific/south-east-asia/asean-china-code-of-conduct-to-conclude-in-three-years/

91 David Francis, John Hudson, and Dan De Luce, "Will Foreign Aid Get Cut on Trump's Chopping Block?", *Foreign Policy*, 23 November 2016, last accessed 20 September 2020, https://foreignpolicy.com/2016/11/23/will-foreign-aid-get-cut-on-trumps-chopping-block/

92 TIME transcript, "Here's Donald Trump's Presidential Announcement Speech", *TIME*, 16 June 2015, last accessed 20 September 2020, https://time.com/3923128/donald-trump-announcement-speech/

93 Go Yamada and Stefania Palma, "Is China's Belt and Road Working? A Progress Report from Eight Countries", *Nikkei Asian Review*, 28 March 2018, last accessed 20 September 2020, https://asia.nikkei.com/Spotlight/The-Big-Story/Is-China-s-Belt-and-Road-working-A-progress-report-from-eight-countries

94 ISEAS–Yusof Ishak Institute report, "The State of Southeast Asia 2020", *ASEAN Studies Centre–Yusof Ishak Institute*, 16 January 2020, p. 36, last accessed 21 September 2020, www.iseas.edu.sg/wp-content/uploads/pdfs/TheStateofSEA-SurveyReport_2020.pdf

95 State Department report, "A Free and Open Indo-Pacific – Advancing a Shared Vision", *The US Department of State*, 4 November 2019, p. 21, last accessed 20 September 2020, www.state.gov/wp-content/uploads/2019/11/Free-and-Open-Indo-Pacific-4Nov2019.pdf

96 *Quoted in* PTI report, "US, China in APEC Cruise Ship Clash on Trade, Influence", *Outlook India*, 17 November 2018, last accessed 20 September 2020, www.outlookindia.com/newsscroll/us-china-in-apec-cruise-ship-clash-on-trade-influence/1423366

97 115th US Congress, "H.R. 5105: BUILD Act of 2018", *The 115th US Congress*, 18 July 2018, last accessed 20 September 2020, www.govtrack.us/congress/bills/115/hr5105/text

98 CRS report, "BUILD Act: Frequently Asked Questions About the New U.S. International Development Finance Corporation", *Congressional Research Service*, 15 January 2019, last accessed 20 September 2020, https://fas.org/sgp/crs/misc/R45461.pdf

99 Glenn Thrush, "Trump Embraces Foreign Aid to Counter China's Global Influence", *The New York Times*, 14 October 2018, last accessed 20 September 2020, www.nytimes.com/2018/10/14/world/asia/donald-trump-foreign-aid-bill.

html?emc=edit_nn_20181015&nl=morning-briefing&nlid=1418504420181015&te=1

100 CRS report, "BUILD Act: Frequently Asked Questions About the New U.S. International Development Finance Corporation", *Congressional Research Service*, 15 January 2019, last accessed 20 September 2020, https://fas.org/sgp/crs/misc/R45461.pdf

101 Adva Saldinger, "How Policy Wonks, Politicos, and a Conservative Republican Remade US Aid", *Devex*, 05 December 2018, last accessed 20 September 2020, www.devex.com/news/how-policy-wonks-politicos-and-a-conservative-republican-remade-us-aid-93934

102 *Quoted in* Glenn Thrush, "Trump Embraces Foreign Aid to Counter China's Global Influence", *The New York Times*, 14 October 2018, last accessed 20 September 2020, www.nytimes.com/2018/10/14/world/asia/donald-trump-foreign-aid-bill.html?emc=edit_nn_20181015&nl=morning-briefing&nlid=1418504420181015&te=1

103 *Quoted in* EXIM press release, "Administration Officials, Congressional and Business Leaders Call for EXIM Board Quorum and Reauthorization", *Export-Import Bank of the United States*, 28 March 2019, last accessed 20 September 2020, www.exim.gov/news/officials-congressional-and-business-leaders-call-for-exim-board-quorum-and

104 Daniel Runde, "Stopping the Buck: Reforming the EXIM Bank So That It Can Finally Be Reauthorized", *The Hill*, 22 May 2019, last accessed 20 September 2020, https://thehill.com/opinion/finance/444776-stopping-the-buck-reforming-the-exim-bank-so-that-it-can-finally-be

105 State Department report, "A Free and Open Indo-Pacific – Advancing a Shared Vision", *The US Department of State*, 04 November 2019, p. 14, last accessed 20 September 2020, www.state.gov/wp-content/uploads/2019/11/Free-and-Open-Indo-Pacific-4Nov2019.pdf

106 State Department report, "A Free and Open Indo-Pacific – Advancing a Shared Vision", *The US Department of State*, 04 November 2019, p. 13, last accessed 20 September 2020, www.state.gov/wp-content/uploads/2019/11/Free-and-Open-Indo-Pacific-4Nov2019.pdf

107 State Department report, "A Free and Open Indo-Pacific – Advancing a Shared Vision", *The US Department of State*, 04 November 2019, p. 16, last accessed 20 September 2020, www.state.gov/wp-content/uploads/2019/11/Free-and-Open-Indo-Pacific-4Nov2019.pdf

108 State Department factsheet, "2019 Indo-Pacific Business Forum Showcases High-Standard U.S. Investment", *The US Department of State*, 03 November 2019, last accessed 20 September 2020, www.state.gov/2019-indo-pacific-business-forum-showcases-high-standard-u-s-investment/

109 Mike Pompeo, "Sec. Pompeo Remarks on 'America's Indo-Pacific Economic Vision'", *US Mission to ASEAN*, 30 July 2018, last accessed 20 September 2020, https://asean.usmission.gov/sec-pompeo-remarks-on-americas-indo-pacific-economic-vision/

110 State Department factsheet, "2019 Indo-Pacific Business Forum Showcases High-standard U.S. Investment", *The US Department of State*, 03 November 2019, last accessed 20 September 2020, www.state.gov/2019-indo-pacific-business-forum-showcases-high-standard-u-s-investment/

111 State Department factsheet, "2019 Indo-Pacific Business Forum Showcases High-standard U.S. Investment", *The US Department of State*, 03 November 2019, last accessed 20 September 2020, www.state.gov/2019-indo-pacific-business-forum-showcases-high-standard-u-s-investment/

112 WH factsheet, "President Donald J. Trump Is Unleashing American Energy Dominance", *The White House – Donald J Trump Administration*, 14 May 2019,

last accessed 20 September 2020, www.whitehouse.gov/briefings-statements/president-donald-j-trump-unleashing-american-energy-dominance/

113 State Department report, "A Free and Open Indo-Pacific – Advancing a Shared Vision", *The US Department of State*, 04 November 2019, p. 17, last accessed 20 September 2020, www.state.gov/wp-content/uploads/2019/11/Free-and-Open-Indo-Pacific-4Nov2019.pdf

114 State Department report, "A Free and Open Indo-Pacific – Advancing a Shared Vision", *The US Department of State*, 04 November 2019, p. 17, last accessed 20 September 2020, www.state.gov/wp-content/uploads/2019/11/Free-and-Open-Indo-Pacific-4Nov2019.pdf

115 Asia EDGE, "Asia EDGE – Enhancing Development and Growth through Energy", *The US Department of State*, n.d., last accessed 20 September, 2020, www.state.gov/asia-edge/

116 State Department report, "A Free and Open Indo-Pacific – Advancing a Shared Vision", *The US Department of State*, 04 November 2019, p. 17, last accessed 20 September 2020, www.state.gov/wp-content/uploads/2019/11/Free-and-Open-Indo-Pacific-4Nov2019.pdf

117 State Department factsheet, "2019 Indo-Pacific Business Forum Showcases High-Standard U.S. Investment", *The US Department of State*, 03 November 2019, last accessed 20 September 2020, www.state.gov/2019-indo-pacific-business-forum-showcases-high-standard-u-s-investment/

118 State Department factsheet, "2019 Indo-Pacific Business Forum Showcases High-standard U.S. Investment", *The US Department of State*, 03 November 2019, last accessed 20 September 2020, www.state.gov/2019-indo-pacific-business-forum-showcases-high-standard-u-s-investment/

119 State Department report, "A Free and Open Indo-Pacific – Advancing a Shared Vision", *The US Department of State*, 04 November 2019, p. 14, last accessed 20 September 2020, www.state.gov/wp-content/uploads/2019/11/Free-and-Open-Indo-Pacific-4Nov2019.pdf

120 State Department factsheet, "2019 Indo-Pacific Business Forum Showcases High-standard U.S. Investment", *The US Department of State*, 03 November 2019, last accessed 20 September 2020, www.state.gov/2019-indo-pacific-business-forum-showcases-high-standard-u-s-investment/

121 State Department factsheet, "2019 Indo-Pacific Business Forum Showcases High-standard U.S. Investment", *The US Department of State*, 03 November 2019, last accessed 20 September 2020, www.state.gov/2019-indo-pacific-business-forum-showcases-high-standard-u-s-investment/

122 David Hutt, "US, Vietnam Ties Have Never Been Better", *Asia Times*, 13 July 2020, last accessed 20 September 2020, https://asiatimes.com/2020/07/us-vietnam-ties-have-never-been-better/

123 Kavi Chongkittavorn, "Can Errant Trump Save ASEAN–US Ties?", *Bangkok Post*, 12 November 2019, last accessed 20 September 2020, www.bangkokpost.com/opinion/opinion/1792349/can-errant-trump-save-asean-us-ties-

124 Reuters report, "U.S. Says Completes Second Aircraft Carrier Visit to Vietnam", *Reuters*, 11 March 2020, last accessed 20 September 2020, www.reuters.com/article/us-vietnam-usa/u-s-says-completes-second-aircraft-carrier-visit-to-vietnam-idUSKBN20Y0F3

125 AFP report, "US Gives Vietnam More Patrol Boats as Defence Links Grow", *Business Standard*, 01 April 2019, last accessed 20 September 2020, www.business-standard.com/article/pti-stories/us-gives-vietnam-more-patrol-boats-as-defence-links-grow-119040100806_1.html

126 Linh Pham, "US to Equip Another Ship for Vietnam Coast Guard", *Hanoi Times*, 21 November 2019, last accessed 20 September 2020, http://hanoitimes.vn/us-to-equip-more-ship-for-vietnam-coast-guard-300241.html

127 Jordan Weissmann, "Who Won America's Trade War with China? Vietnam", *Slate*, 05 February 2020, last accessed 20 September 2020, https://slate.com/business/2020/02/trade-war-china-us-vietnam-trump.html

128 David Hutt, "US, Vietnam Ties Have Never Been Better", *Asia Times*, 13 July 2020, last accessed 20 September 2020, https://asiatimes.com/2020/07/us-vietnam-ties-have-never-been-better/

129 State Department report, "A Free and Open Indo-Pacific – Advancing a Shared Vision", *The US Department of State*, 04 November 2019, p. 13, last accessed 20 September 2020, www.state.gov/wp-content/uploads/2019/11/Free-and-Open-Indo-Pacific-4Nov2019.pdf

130 Nhan Dan report, "US, Vietnam Sign US$42 Million Deal to Enhance Vietnam's Economic Competitiveness", *Nhân Dân*, 15 April 2020, last accessed 20 September 2020, https://en.nhandan.org.vn/business/item/8577202-us-vietnam-sign-us$42-million-deal-to-enhance-vietnam's-economic-competitiveness.html

131 Hindu editorial, "Sanctions Relief: On Waiver for India Under CAATSA", *The Hindu*, 26 July 2018, last accessed 20 September 2020, www.thehindu.com/opinion/editorial/sanctions-relief/article24514443.ece

132 Pizaro Gozali Idrus and Dandy Koswaraputra, "Indonesia to Host US Companies Relocating from China", *Anadolu Agency*, 16 May 2020, last accessed 20 September 2020, www.aa.com.tr/en/asia-pacific/indonesia-to-host-us-companies-relocating-from-china/1843244

133 Charissa Yong, "PM Lee, Trump Renew Key Defence Pact on US Use of Singapore Air, Naval Bases", *The Straits Times*, 24 September 2019, last accessed 20 September 2020, www.straitstimes.com/world/pm-lee-trump-renew-key-defence-pact-on-us-use-of-singapore-air-naval-bases

134 Valerie Insinna, "Singapore Gets the Green Light to Buy F-35s", *Defense News*, 09 January 2020, last accessed 20 September 2020, www.defensenews.com/air/2020/01/09/singapore-gets-the-green-light-to-buy-f-35s/

4

TRUMP AND SOUTH ASIA

The Trump administration's policy towards the Indo-Pacific region encompassed a degree of continuity with the predecessor administration's Rebalance to the Asia-Pacific policy. However, at the same time, under its policy of seeking "fair and reciprocal" trade, the Trump administration departed from the US policy precedent of compartmentalising American defence commitments toward allied and partner nations. With allies like Japan and South Korea, Trump sought to exact renewed trading arrangements or equitable burden-sharing by putting US security commitments in a state of limbo.

In case of nascent American partners, the Trump administration's realpolitik approach towards traditional American allies in the region encouraged their independent capacity-building – surely with American assistance but *not* its largesse. This design in the Trump administration's approach of cultivating partners – and not dependents – is most apparent with respect to the South Asian subregion of the Indo-Pacific construct. For instance, the Trump administration recognised the US security establishment's long-standing view on the interlinked destinies of the Indian and Pacific oceans, by elevating South Asia in the Indo-Pacific strategy. In this way, the US also recognised India's maritime providence over the Indian Ocean region, seemingly deferred to India's "strategic autonomy" over charting its rise on account of its own interests, and embraced complementarities with India's 'Act East' policy.

Since the Second World War, American primacy has been undergirded by its impressive power projection capabilities. Throughout the Cold War and thereafter, those capabilities have been apparent with the US maintaining a constellation of 'hub & spokes'[1] partnerships and multilateral alliances in order to sustain American operational advantage towards contingencies in almost every major flashpoint region of the world. The result has been the US forging defence partnerships with about 69 countries – which account for over 75 percent of the global economic output and are home to more than 2 billion people,[2] and extensive power projection capabilities via nearly 800 US bases and outposts in over 70 countries around the world.[3] This extensive US military presence was carved into half a dozen combatant commands with specified geographic purviews. In Asia, India – hardly a Pacific power – fell under the purview of the Pacific Command (USPACOM), whereas its South Asian neighbours Pakistan and Afghanistan were clubbed with the Middle East and North Africa (MENA) region under the US Central Command (USCENTCOM).[4]

Probably reflective of the US security community's foresight on India's potentialities in its eastward security context, an American political consensus on courting India's rise as a "counterweight" to China's ascent came to the fore only in the post–Cold War era under the George W Bush administration. Although it is important to note that binary, *Manichaean* perspectives on China and India figured in American political discourse even in the heydays of the Cold War (e.g. Sen. John F. Kennedy's 1959 speech on "Red" China and "Free" India).[5] However, those conceptions were short-lived in view of India's gradual tilt towards the Soviet Union, the US' rapprochement with China, and centrality of the European theatre in the US–USSR Cold War.

With the Trump administration's rechristening of USPACOM in Hawaii to the Indo-Pacific Command, however, the US accorded policy heft to the American military's operational purview of the region and concurrently recognised the centrality of South Asia – and by that extension India. As for the other connotation of the early operational allocation of India in the Pacific context, the USPACOM-USCENTCOM divide precariously fell along the India–Pakistan border to denote the de-hyphenation of India and Pakistan in the US' strategic calculus. The Trump administration initially struggled with continuing this policy precedent.

The ebbs and flows of Trump's perspective on India–Pakistan–Afghanistan

In the post–Cold War timeline, the United States' policy towards South Asia has primarily been riddled with hurdles posed by the historical rivalry between India and Pakistan. Soon after the collapse of the Soviet Union, South Asia emerged as a flashpoint with nuclear tests from both India and Pakistan, and the subsequent military confrontation between them in the Kargil district of Kashmir. This prompted then-US President Bill Clinton to deem the Indian subcontinent and the line of control in Kashmir as the "the most dangerous place in the world".[6]

Subsequently, under the Clinton administration's grand strategy of "enlargement", i.e. prioritising newer American partnerships with like-minded democracies,[7] Clinton came to hone a positive outlook towards New Delhi and even paved the way for off-ramping US sanctions. Thereafter, US courtship of India also assumed a bipartisan fervour with the two nations' shared experience with the menace of terrorism – with the attacks on the Indian Parliament closely following the attacks on September 11, 2001 (9/11). With added policy momentum, the US under Republican President George W Bush even sought to cultivate India as a strategic partner with overtures like recognising its nuclear status with the India–United States civil nuclear agreement.

Although 9/11 proved to be instrumental in forging an American political consensus on courting India, the US' response to the terror attacks presented a challenge. With the US military's massive regime-change undertaking in Afghanistan under the Global War on Terror (GWOT), the US' effort to dampen perceptions of its Cold War–era proximity towards India's rival – Pakistan – stood undercut. As the war effort in Afghanistan dragged on – from the initially intended regime change operation to overthrow the Taliban to a comprehensive counterinsurgency operation against a mosaic of terror outfits operating along the Afghanistan–Pakistan border – the US' operational dependency on Pakistan grew. This figured in terms of increased US aid for Pakistan, for instance, with Islamabad receiving nearly US$ 30 billion in civilian aid and military reimbursements from Washington since 2001.[8] Most importantly, this concurrently posed hurdles in the evolving US–India dynamic, such as in the case of America's slow-walked, largely symbolic actions in support of India's efforts against cross-border terrorism. Case in point: only ahead of US President Barack Obama's maiden visit to India in 2010 did the US Department of Treasury designate and sanction Azam Cheema as an operative of the Pakistan-based Lashkar-e-Taiba for his role as the operational commander of the 2008 Mumbai attacks.[9]

Hence, in ascertaining Trump's record on balancing the US' operational dependence on Pakistan with its cultivation of India as a strategic partner, the relevance of his administration's policy towards Afghanistan cannot be ignored.

Breaking the policy inertia on Afghanistan

In 2016, with its conservative nationalist bent against American internationalism, the Trump campaign rallied against the Washington elites' record on "endless wars". Trump criticised the needless squandering of American resources – both in treasure and blood, at the cost of neglecting priorities at home. One such instance he continually invoked was the US war effort in Afghanistan, which by then was already in its 17th year and steadily inching to surpass the US' involvement in Vietnam (which lasted 17 years and four months) as America's "longest war".

Upon taking office, Trump ordered a "comprehensive review" on the US' strategy in Afghanistan and then announced his administration's South Asia policy.

In an address on August 21, 2017,[10] Trump narrowed America's aim in Afghanistan by criticising the US' focus on nation-building. On the US' failures, Trump rallied against Pakistan by stating the US could "no longer be silent about Pakistan's safe havens for terrorist organizations, the Taliban, and other groups that pose a threat to the region and beyond". He added:

> Pakistan has also sheltered the same organizations that try every single day to kill our people. We have been paying Pakistan billions and billions of dollars at the same time they are housing the very terrorists that we are fighting. But that will have to change, and that will change immediately. No partnership can survive a country's harboring of militants and terrorists who target U.S. service members and officials. It is time for Pakistan to demonstrate its commitment to civilization, order, and to peace.

Thereafter, in exacting a commitment towards "decisive and irreversible action" against "externally-focused militant groups and UN-designated terrorist organizations operating from its territory", the Trump administration followed through over the next two years by withholding US aid from Pakistan.[11] In a tweet, he said, "The United States has foolishly given Pakistan more than 33 billion dollars in aid over the last 15 years, and they have given us nothing but lies & deceit, thinking of our leaders as fools. They give safe haven to the terrorists we hunt in Afghanistan, with little help. No more!"[12] Starting in mid-2017, the Trump administration announced a "pause" in the disbursement of US$ 255 million in foreign military financing (FMF).[13] It later followed up with broader security aid suspension in January 2018 – which affected the US' annual US$ 1.3 billion security assistance to Pakistan – and in mid-2019 slashed the US' commitment under the Pakistan Enhanced Partnership Agreement (PEPA) by US$ 440 million.[14] In all, Pakistan was reported to have lost US$ 1.66 billion in security assistance initially,[15] and another "US$ 790 million in unobligated" foreign military financing (dating back to FY2001) was held back.[16] This was also coupled with action at multilateral fora, with the US supporting Pakistan's greylisting at the Financial Action Task Force (FATF) for "outstanding counterterrorism deficiencies".[17]

Meanwhile, on Afghanistan, Trump altered the US' approach towards gaining an on-ground initiative and coax the Taliban to the negotiating table. In announcing "conditions on the ground – not arbitrary timetables" as the guiding approach,[18] the Trump administration sanctioned a troop surge, accorded greater operational latitude to US military leaders, and increased the frequency of US airstrikes. In an approach reminiscent of Richard Nixon's in Vietnam, the Trump administration's effort to put the US on a "path to win" included:[19] approving a surge of nearly 7,000 troops; dropping the US' most powerful nonnuclear bomb on a cave complex in eastern Nangarhar province (i.e. the MOAB) in April 2017; overseeing the use of more munitions in Afghanistan "in 2019 than any other year since at least 2010" and US forces conducting strikes in "27 of Afghanistan's 34 provinces in the first two months of 2020 alone";[20] and, finally, permitting US diplomats to engage directly with the Taliban by appointing noted US diplomat Zalmay Khalilzad as Special Representative for Afghanistan Reconciliation.

However, this attempt at breaking the policy inertia by adopting economic coercion to coax Pakistan to fulfil its obligations towards the US and permitting an intensified military campaign to aim for an on-ground initiative brought the US back to square one.

Pakistan's resurgence in the US' Afghan calculus

Two years into the Trump presidency, it became apparent that "turning corners" in Afghanistan was a confounding effort. Just as the Obama administration's own experience with a military-intensive solution (following the 2009 review of its Afghanistan policy), Trump's off-handed approach with the military effort in Afghanistan produced little results.

Apart from the intensification of airstrikes, the Trump administration had at one point even deferred to then-Secretary of Defence James Mattis's judgement on setting troop levels and supported the Pentagon's information blackout – with respect to not revealing active troop numbers and civilian casualties.[21] Furthermore, the Special Inspector General for Afghanistan Reconstruction (SIGAR) – the primary US government oversight authority on Afghanistan reconstruction – noted that the US military had stopped "producing its district-level stability assessments of Afghan government and insurgent control and influence" because apparently it was deemed to be "of limited decision-making value to the [U.S.] Commander".[22] As per the last reported metrics in early 2019, the share of districts "under government control or influence fell to 53.8%, as of October 2018. This figure was the lowest recorded by SIGAR since tracking began in November 2015; 12% of districts were under insurgent control or influence, with the remaining 34% contested".[23] Moreover, the militarised approach also spurred a sharp increase in civilian casualties. According to a United Nations report, the third quarter of 2019 saw the highest number of civilian casualties – over 4,300 killed and injured, since it began tracking Afghanistan.[24]

This prompted a gradual increase in Pakistan's strategic currency with the United States, in line with its own interests against an ethnic Pashtun-dominated civilian government in Kabul which could, over time, enflame its own large Pashtun minority. To this end, Pakistan's role in the Afghan conflict pertained to its ties with the Taliban – which it also views as "a relatively friendly and reliably anti-India element in Afghanistan"[25] – and high-profile affiliates like the Haqqani Network.

In 2018, as the writing-on-the-wall over the inefficacy of the US' militarised approach in Afghanistan became apparent, Pentagon's civilian leadership and senior military officials like Gen. Joseph Dunford, then-chairman of the Joint Chiefs of Staff, described the situation in Afghanistan as a "stalemate" on multiple occasions.[26] On the diplomatic front, the administration's decision to permit direct talks with the Taliban also meant a significant departure from US policy precedent of prioritising an "Afghan-led, Afghan-owned reconciliation process". In then courting Pakistani assistance, Trump reconfigured US aims – from "winning" to heading for the exits.

Most notably, during an Oval Office meeting (in July 2019) with Pakistani Prime Minister Imran Khan, Trump said:

> I think Pakistan is going to help us out to extricate ourselves. We're like policemen. We're not fighting the war. If we wanted to fight a war in Afghanistan and win it, I could win that war in a week. I just don't want to kill 10 million people. Does that make sense to you? I don't want to kill 10 million people. I have plans on Afghanistan that, if I wanted to win that war, Afghanistan would be wiped off the face of the Earth. It would be gone. It would be over in – literally, in 10 days. And I don't want to do – I don't want to go that route.[27]

The elevation of Pakistan's role in the US' attempt to "extricate" itself, was already evident since late 2018. Upon hitting multiple roadblocks with the Taliban on the US' demand for a ceasefire as a precondition for talks in order to establish a precedent of either sides' sincerity on negotiations, the US invited Pakistani assistance. In October 2018, for instance, the US sought the release of Taliban co-founder Mullah Abdul Ghani Baradar from Pakistani custody. In early 2019, as two weeks of US–Taliban negotiations in Doha, Qatar, ended with Khalilzad announcing "a withdrawal timeline and effective counterterrorism measures" to have been "finalized", he revealed the criticality of Pakistan's actions.[28]

Announcing Baradar's release to have been at the behest of the US, he also thanked Pakistan for its support in facilitating the Taliban delegation's travel to Doha.[29] According to reports, Baradar's presence helped in alleviating a persisting "environment of mistrust where each side believed the other could be using negotiations as a distraction to gain advantage on the battlefield" while keeping "the talks going".[30] It was also reported to be instrumental in lending "authority to the Taliban's negotiating team that was lacking in the past, with negotiators frequently having to wait for answers from the leadership in Pakistan".[31] Most crucially, it was Baradar that went on to sign the US–Taliban agreement alongside Khalilzad in 2020. The finalised deal included assurances from the Taliban on not permitting Afghan territory to be used for plotting attacks in exchange for the US reducing its forces in the first 135 days of the deal down to 8,600 (i.e. back to pre-Trump era numbers) and thereafter removing all troops within 14 months.[32] Given the latter aspect's relevance also to President Trump's 2016 campaign pledge of ending the US military's effort in Afghanistan, a concurrent dampening in rhetoric towards Pakistan was apparent.

The January 2020 iteration of the Pentagon's biannual report on Afghanistan highlighted Pakistan's role in "supporting the Afghan reconciliation" and described its role as "constructive but limited".[33] Secretary of State

Mike Pompeo lauded Pakistan's "sincere efforts" towards Afghan reconciliation and the peace process.[34] Ahead of the February 2020 FATF meet on deciding Pakistan's further demotion from the grey-list to the blacklist, US Assistant Secretary of State for South and Central Asia Alice Wells said, "It is heartening to see that Pakistan's government has made significant progress on these matters and that too in a short time".[35] Over the course of the US–Taliban negotiations, this also translated into shifts beyond softer US rhetoric. For instance, the US approved a US$ 125 million "support program for Pakistan's fleet of F-16 warplanes"[36] and announced its "intent to allow Pakistan to rejoin" the International Military Education and Training (IMET) for military cooperation "following a two-year suspension".[37]

This resurgent relevance of Pakistan in the US' Afghan calculus fed the Trump administration's momentary inconsistency on the de-hyphenation of India and Pakistan.

Risking de-hyphenation of India–Pakistan

Much like Trump, upon assuming office in 2009, Barack Obama also conducted an administration-wide review of US policy in Afghanistan and South Asia at large. Motivated by his own 2008 election campaign promise of winding down America's protracted wars, the review was geared towards exploring civilian-led options in the otherwise overtly militarised GWOT effort. Having campaigned against the militarisation of US foreign policy, Obama tapped veteran US diplomat Richard Holbrooke as a key facilitator of the administration's interagency search for solutions. Holbrooke had made his bones in the 1990s with his success in brokering peace between warring factions in Bosnia, which culminated with the Dayton Peace Accords under Bill Clinton.[38]

Subsequently, in the released nine-page White Paper on the Interagency Policy Group's Report on US Policy toward Afghanistan and Pakistan, the Obama White House presented options to further the US' insistence on an Afghan-led reconciliation process, purported a narrow threat conception limited to targeting al Qaeda fighters and its infrastructure (instead of an all-out expansive approach against the Taliban), emphasised the use of developmental assistance as a chief policy tool towards strengthening the civilian dispensation in Afghanistan and Pakistan, and warned against a commitment spiral spurring counterintuitive results.[39]

However, amidst the inter-agency process, a significant divergence emerged with respect to Obama and Holbrooke's fundamental perceptions of the region. Holbrooke sought to instil an integrative US approach which recognised Pakistan as the "center stage" to the "sideshow" of the US' precipitous efforts in neighbouring Afghanistan.[40] In viewing Afghan stability as being inadvertently linked to Pakistani stability, Holbrooke reportedly believed that "a crucial step to reducing radicalism in Pakistan was to ease the Kashmir dispute with India, and he favored more pressure on India to achieve that".[41]

Obama's subsequent dismissal of this hyphenation, however, stemmed from two factors.

First, the US president found himself boxed-in at the hands of the US military's preference for an expansive conception of US interests in Afghanistan. Then-Commander of the International Security Assistance Force (ISAF) General Stanley McChrystal pushed his expansive assessment of the war effort via the media.[42] With unusual media appearances by McChrystal and his nearly 70-page "COMISAF Initial Assessment" draft mysteriously finding its way into the hands of influential journalist Bob Woodward at *The Washington Post*,[43] his preference for a continued counterinsurgency effort in Afghanistan limited Obama's capital on pursuing a different approach. In face of warnings of "mission failure" in the "COMISAF Initial Assessment", Obama's grudging approval of a troop surge of 30,000 troops (albeit with a predetermined deadline of July 2011) signified a return to the erstwhile US military-intensive, nation-building approach.[44]

Second, in the post–Cold War timeline, there had been an American policy precedent of de-hyphenating India from Pakistan, towards courting the former as a long-term strategic partner. For instance, foreseeing the impediments posed by the historical rivalry between India and Pakistan, the Clinton administration sought to compartmentalise its approach on off-ramping sanctions against the two South Asian nations. Despite both countries declaring themselves as nuclear weapon states in mid-1998, the Clinton administration sought an approach towards India and Pakistan which would be independent of each other. Thus, even as the Clinton administration excluded agricultural exports from the purview of sanctions for both, it continued to support India's developmental goals with the US supporting World Bank loans towards child education and healthcare projects for instance.[45] The Bush administration continued this precedent with US military aid for Pakistan continuing to remain frozen until the post-9/11 war in Afghanistan warranted a shift in policy. And most crucially, the Bush administration (via the India–United States civil nuclear agreement) only accorded India with international recognition of its nuclear status.

Thus, in continuing this precedent of two predecessor administrations, Obama acted upon Holbrooke's prescription only to the extent of recognising the interconnectedness of Afghan and Pakistani stability, with actions such as institutionalising Holbrooke's thinking in his designated purview as the US Special Representative for Afghanistan *and* Pakistan (SRAP). Moreover, the prospect of India being included under SRAP's purview was met with strong protest from New Delhi. Notably, then-External Affairs Minister Pranab Mukherjee met then-US Ambassador David Mulford and said the consideration "smacks of interference and would be unacceptable".[46]

Trump, however, seemed to be unbeknownst about the de-hyphenation policy's core prescription of the US not taking sides over the Kashmir conflict and emphasising the "pace, scope, and character" of any resolution of the matter to solely be limited to the two parties.[47]

For instance, following the discussed early 2019 US–Taliban meeting, the July 2019 Oval Office meeting with Imran Khan witnessed Trump seemingly complementing Pakistani apprehensions on Kashmir by claiming Prime Minster Narendra Modi to have asked if he'd like to be a "mediator or arbitrator" on the matter. In responding to Khan's comments at the meeting, Trump said: "If I can help, I would love to be a mediator".[48] In terms of optics, this not only came across as the US reciprocating on Pakistan's long-standing push to invoke external arbitration but also struck at the heart of India's well declared position against internationalising the Kashmir matter. In dampening the fallout over Trump's offer, American legislators sought to contain any prospective impact on the broader US–India bilateral trajectory. For instance, co-chairs of the Congressional Caucus on India and Indian Americans in the US House of Representatives offered a rebuke of the US president's offer. In a bipartisan joint statement, Rep. George Holding and Rep. Brad Sherman clarified: "Consistent with decades of US policy, we believe the dispute over Kashmir must be resolved bilaterally by India and Pakistan. The Republic of India is one of America's closest and most important allies, and we look forward to working with Prime Minister Modi and Indian officials to combat terrorism and extremism throughout the region".[49]

In another instance, following India's abrogation of Article 370 – which ended Kashmir's special status – Trump repeated the offer to mediate all through August and September 2019. However, the sincerity of his offer stood undercut with back-to-back instances, wherein Trump was seen expediently supporting the party sitting across him. On August 26, 2019, alongside Modi at the G7 summit in Biarritz (France), Trump said, "We spoke last night about Kashmir . . . and the Prime Minister [Modi] really feels he has it under control. We spoke about it last night at great length".[50] Then, on September 23, 2019, on the sidelines of the UNGA with Imran Khan sitting beside him, Trump said, "I am ready, willing and able. It's a complex issue and it's been going on for a long time. But if they are both willing then I'm ready to do it".[51] Similarly, Trump made another offer ahead of the February 2020 signing of the US–Taliban deal. At a media interaction ahead of his meeting with Imran Khan in Davos, Trump said, "We can meditate on the issue. We are working with Pakistan on some borders and we were talking about Kashmir in relation to what is going on between Pakistan and India. If we can help, we certainly will be helping. We have been watching that and following it very very closely".[52] Eventually, the Trump administration backed off from this proposal and sustained a position in lockstep with India's stance against internationalising the Kashmir matter and deemed the abrogation of Article 370 as India's "internal matter".[53]

Through this period, wherein the de-hyphenation of India and Pakistan seemed to be under threat, the Trump administration largely continued its policy of maintaining pressure on Pakistan for its lacklustre record on supporting the US' counterterrorism efforts. However, the resurgent traces – albeit mostly in rhetoric – of the Holbrooke thinking in the US' approach did momentarily risk raiding the trajectory of US–India ties – which under Trump witnessed its own set of challenges.

Trump's upending of the Carter mantra

Upon assuming office, Trump oversaw the prioritisation of seeking the recalibration of America's trading relationship. In three years, Trump exacted renewed trading arrangements with Canada and Mexico after labelling them as "threats" to national security, negotiated trade deals with Japan and South Korea after briefly linking US security commitments to American apprehensions on trade, and scored a limited Phase One deal with China after a nearly 18-month-long "tech-economic" conflagration. Thus, in exacting "fair and reciprocal" trading arrangements, Trump also turned his ire towards India.

The first US action came in March 2018, when the Trump administration – in its bid to arrest the decline of America's heavy metal industries – levied tariffs of 25 percent and 10 percent on steel and aluminium, respectively. By invoking Section 232 of the Trade Expansion Act of 1922, the Trump administration construed US import of steel and aluminium to have "impaired national security".[54] Although the administration's central concern was the influx of surplus Chinese steel and aluminium in the global markets, the action also included India, which already held a very limited share of the American market. In possibly recognising this move to have mostly stemmed from the mentioned China factor, India at that point did not impose retaliatory tariffs in "hopes of averting a trade war and bridging economic differences".[55]

Until then, India had found itself in the Trump administration's crosshairs, but mostly in rhetoric. For example, during a February 2018 meeting with US governors, Trump referenced a conversation with Modi. Trump recounted: "When Harley Davidson sends a motorcycle to India, as an example, they have to pay 100 percent tax. . . . Now, the Prime Minister, who I think is a fantastic man, called me the other day and he said we are lowering it to 50 percent. I said okay, but so far, we're getting nothing. So we get nothing, he gets 50 [percent], and they think like they're doing us a favour. That's not a favour".[56]

Ahead of the 2019 general elections in India, however, tensions escalated beyond rhetoric. The US announced its intent to end India's status under the Generalized System of Preferences (GSP) – a preferential trading arrangement unilaterally offered by the US to encourage free trade with developing nations.[57] India had been the biggest beneficiary of the programme with its exports to the US accounting for "over a quarter of the goods that got duty-free access into the US in 2017".[58] Moreover, about 12 percent (worth US$ 5.58 billion) of all Indian exports to the US in 2017 benefitted under the GSP scheme.[59] In foreseeing this move's probable impact on the general elections in India, American legislators of the Congressional Caucus on India and Indian Americans and the Senate India Caucus urged the office of the US Trade Representative (USTR) – and the Trump administration by extension – to delay any decision on the matter. The co-chair of the Congressional Caucus on India and Indian Americans, Rep. George Holding wrote: "It is my request that the administration postpone the termination of India's GSP eligibility and revisit this decision after India's general election. At that point, we would be free from the turmoil of politics and would be able to have a more productive conversation".[60] After the elections concluded – with Modi returning to power with a larger mandate, the USTR terminated India's status under the GSP programme on account of "its failure to provide the United States with assurances that it will provide equitable and reasonable access to its markets in numerous sectors" and owing to India no longer qualifying for the "statutory eligibility criteria" as a developing nation.[61] At this point, New Delhi responded with duties on 28 US goods. For instance, duty on US walnuts went up to 120 percent and to 70 percent for chickpeas.[62]

Until the termination of India's GSP benefits, trade frictions were seen as a matter of Trump's 'America First' worldview clashing with Modi's 'Make in India' initiative. However, data reflected a positive trend on the US–India trade imbalance, with it gradually correcting itself on account of rising US exports to India. According to the USTR's own figures,[63] US–India goods and services trade reached "an estimated $142.6 billion in 2018", wherein US' exports and imports stood at US$ 58.7 billion and US$ 83.9 billion, respectively. Goods exported to India stood at "$33.5 billion, up 30.6% ($7.9 billion) from 2017 and up 89.5% from 2008", whereas services export was "an estimated $25.2 billion in 2018, 6.6% ($1.6 billion) more than 2017, and 151% greater than 2008 levels".

This meant, from 2017 to 2018, America's trade deficit with India (already under a paltry US$ 30 billion) further decreased by 9 percent for goods and by 3.7 percent for services.[64] However, the termination of GSP benefits and the concurrent conception of India as a 'developed nation' (despite its per capita gross national income being under $12,375, which would render it as a lower-middle-income economy)[65] reflected the Trump administration's intent to address the US' long-standing apprehensions over India's market-access barriers. This meant an upending of the Carter mantra.

In a bid to continue the Bush-era thaw in US–India ties, the Obama administration's outlook towards India was shaped by the country's long-term strategic potentialities with respect to the US' evolving strategic competition with China. Postulated by Obama's last secretary of defense, Ashton Carter, the unofficial dictum of the Carter mantra informed the administration's focus on US–India strategic ties, all whilst preventing differences (such as that over market access) from crowding out minimal-yet-positive developments.[66]

The Trump administration, however, sought to play out differences in the open and even contemplated escalatory measures against India. For instance, the Trump administration reportedly contemplated limiting H1B visas to 15 percent for any country like India that insists on data localisation.[67] Moreover, in a sign of also expanding the scope of apprehensions, US highlighted India's restrictions on cross-border data flows and data localisation requirements as an "onerous" requirement in its 2019 National Trade Estimate.[68]

Such nascent divergences compounded American negotiators' requests for addressing long-standing issues like India's insistence on certifying dairy imports and its price caps on pharmaceutical imports. However, Indian negotiators sought to rationalise their positions. In the case of the former, India has cited cultural sensitivities over the use of blood-meal by US dairy manufactures to attain greater yield from their cattle.[69] In case of the latter, the lowering of prices for coronary stents and knee implants (by 85 percent and 65 percent, respectively)[70] has been attributed to India's attempt to prevent "exorbitant pricing from adversely affecting the Indian consumer, a largely middle-income base".[71]

Moreover, given India's approach to trade negotiations which largely pertained to addressing the trade imbalance between the two countries, attempts to soothe tempers encompassed assuring the furtherance of the discussed trend on the reduction in the US' trade deficit. A near-balanced US–India trade dynamic was assured in the face of India's rising defence imports as the US became India's second-largest arms supplier; future purchases of civil aviation aircraft with India poised to order over 2,300 new planes over the next 20 years; and, finally, increased energy imports with US crude exports to India increasing threefold in early 2019 alone.

However, American apprehensions over divergent outlooks on trade negotiations only exacerbated tensions. For instance, in 2019, the Trump administration reportedly even contemplated a "full-blown [Section 301] investigation" into India's market access barriers at large.[72] Thereafter, that was seemingly postponed in view of reports of a partial US–India trade deal in the works. Ahead of Trump's maiden visit to India in February 2020, however, negotiations even for a "mini-deal" reportedly broke down.[73] Thus, under Trump, US trade apprehensions and negotiations with India dragged on for over two years, only to yield multiple stalemates on account of the two countries' divergent approaches to trade tensions.

Institutionalising US–India strategic ties and mitigating India's westward apprehensions

With Trump's India policy replete with persistent trade tensions, the strategic aspect of the bilateral dynamic underwent a period of consternation.

Most notably, in the early years of the Trump administration, many in the Indian strategic community questioned the wisdom of seeking greater alignment with the US. With American credibility in shambles owing to Trump's brutish behaviour towards some of the US' long-standing allies across Western Europe and Northeast Asia, Indian commentators questioned why New Delhi – a nascent and an informal American partner – would escape the heightened focus on correcting trading arrangements. Moreover, given Trump's upending of the Carter mantra – which discouraged differences over trade and market-access playing out in the open, the punditry stood validated. However, on the other significant facet of the discussed mantra, i.e. furthering strategic ties, the Trump administration surprisingly bore considerable continuity with its predecessor administration.

After going through its own share of ebbs and flows over policy in the Af-Pak theatre impacting the US' strategic outreach to India, the Obama administration put considerable emphasis on institutionalising the bilateral relationship. Without a formalised defence pact or even a trading arrangement, the US–India trajectory was mostly animated with the inertia accorded by convergent strategic interests and the chemistry between top political leadership – which over time translated into a bipartisan fervour on both sides. Since the Bill Clinton and George W Bush administrations began the US' strategic outreach to India at the turn of the century, bilateral trade breached US$ 142.6 billion in 2018.[74] In terms of volume, the same has come to resemble America's trade with allies like South Korea (US$ 167 billion) and France (US$ 129 billion).[75] Over time, India also begun to conduct "more joint military drills, tabletop exercises, and defense dialogues with the US than with any other country, which include more than 50 'cooperative events across all Services' annually".[76]

In foreseeing the need for definite frameworks and standardised communication channels, the Obama administration, in its work with the Manmohan Singh government, pressed for the initiation of consultative ministerial forums. Case in point is the November 2010 announcement of the Homeland Security Dialogue between the Ministry of Home Affairs and the Department of Homeland Security.[77] On force interoperability, the Obama administration in its final year – and even despite a change of guard in India with the 2014 election victory of Narendra Modi – finalised the Logistics Exchange Memorandum of Agreement (LEMOA). The India-specific version of the Acquisition and Cross-Servicing Agreement (ACSA) aimed to facilitate "basic terms, conditions, and procedures for reciprocal provision of logistic support, supplies, and services between the armed forces of India and the United States".[78]

Building on this, one year into the Trump administration, the US and India signed another force interoperability pact – the Communications Compatibility and Security Agreement (COMCASA). An exclusive tailored version of the Communication and Information on Security Memorandum of Agreement (CISMOA), the pact seeks to facilitate "access to advanced defense systems and enable India to optimally utilize its existing US–origin platforms".[79] This was followed by the US and India inking the final interoperability agreement – the Basic Exchange and Cooperation Agreement (BECA), for mutual access of geospatial maps.[80] Under Trump, the two sides also instituted the annual US–India 2+2 consultative dialogue between foreign and defence cabinet chiefs, the India–US Strategic Energy Partnership working group between India's Petroleum Minister and US Energy Secretary, and a hotline between Indian and American National Security Advisors – which proved to be pivotal in aligning Indian and American responses following the February 2019 terror attack in Pulwama.[81]

Furthermore, in according a much needed fillip to US–India defence co-production and co-development, the Trump administration and the Modi government successfully finalised the Industrial Security Annex (ISA). Deemed to be a crucial precursor for the actualisation of the Obama-era US–India Defence Technology and Trade Initiative (DTTI), the ISA seeks to "facilitate the exchange of classified military information between Indian and the US defence industries".[82] In a testament to the efficacy of US–India consultative platforms that reduce the relationship's dependency on chemistry between the heads of state, the two sides signed the ISA at the second iteration of the US–India 2+2 dialogue, after the two sides had expressed "their readiness to begin negotiations" on the matter at the inaugural 2+2 dialogue in 2018.

Apart from these proactive measures towards institutionalising US–India strategic ties, the Trump administration in its work with the Modi government also sought to iron out policy-level divergences in the two nations' views on the Indo-Pacific.

Accord India with eyes at the USCENTCOM

Under Trump, a significant development from the standpoint of institutionalising US–India strategic ties was one which went beyond technical interoperability. In a bid to enhance contact points between Indian and American forces, the Modi government accepted the Trump administration's offer to place a liaison offer at the US Naval Forces Central Command (NAVCENT) – the US Central Command's (USCENTCOM) naval component, which is also known as the home of the US Fifth Fleet. India announced that the defence attaché at its embassy in Bahrain will also serve as India's representative at NAVCENT.[83] This move was significant, as in the past, India had refused the US' offer to post a liaison officer at US Pacific Command (USPACOM) in Hawaii, citing its continued westward security concerns.

In post–Cold War US administrations' bid to have India emerge as a prominent actor in its eastward security calculus – under the Pivot to Asia, Rebalance to Asia, and now the Indo-Pacific strategy – New Delhi's westward apprehensions mainly hindered India's complete integration. In declining multiple offers since 2002, the Atal Bihari Vajpayee government most notably did not reject the offer. Instead, it asked for also posting an officer at CENTCOM primarily due to security threats (mainly cross-border terrorism) emanating from neighbouring Pakistan. However, due to the US' operational dependence on Pakistan under the imperatives of the GWOT effort in Afghanistan, the US balked at India's counter-offer.[84] Under Trump, amidst his administration's pressure campaign in order to exact Pakistani compliance against the Taliban and other terror outfits operating in the Af-Pak theatre, India's request for eyes at USCENTOM may have suddenly become palatable. Following this development which recognised India's westward apprehensions and the Trump administration's renaming of USPACOM to the Indo-Pacific Command in its evident recognition of India's centrality under the Indo-Pacific strategy, the US effectively interlinked the destinies of the Indian and Pacific oceans. As a result, the Modi government is now reportedly also considering US requests to post an Indian naval liaison officer at the American combatant command in Hawaii.[85]

Coerce a shift in India's energy security

Another facet of India's westward focus stemmed from its energy security. As per 2018–19 data, four of India's five biggest sources of crude were Middle Eastern nations – with Iraq topping the list with more than 46.61 million tonnes (MT), Saudi Arabia at 40.33 MT, Iran with 24 MT, and, finally, the UAE at 17.49 MT.[86]

Under the Trump administration's "maximum pressure" policy, US–Iran tensions occasionally flared. With the creeping influence of neoconservative thought towards furthering a 'regime change' agenda,[87] the Trump administration sought to cut the Ayatollah Ali Khamenei regime's prime source of income. Following the US' withdrawal from the Joint Comprehensive Plan of Action (or simply the Iran Nuclear Deal), the Trump administration announced new rounds of sanctions mainly towards ending Iran's revenue from oil, since oil and petroleum products made up "about 70 percent of the country's exports by value".[88]

However, the US initially did so with a 180-day exemption period for India and seven other countries to gradually wean themselves off Iranian crude. During this period, most exempted countries continued to import Iranian oil in view of historical ties – especially India as it enjoyed a "60-day credit period, free insurance, and cheaper freight" on account of its long-standing historical ties with Iran.[89] As a result, Iran's export to India increased by "6.24 per cent to 24 MT in 2018–19".[90] However, at the end of the exemption period in 2019, the Trump administration announced it would not extend waivers in view of limited gains for its policy – in terms of coaxing Iran back to the negotiating table.

Moreover, a transactional approach also came to be associated with the US' attempt to shift the dynamics of India's energy choices. Most notably, ahead of the designation of JeM chief Masood Azhar under the UN 1267 sanctions list, reports emerged about the US suggesting India's compliance on reducing Iranian oil imports in exchange for the Trump administration's efforts at the UN against China's "technical hold" over the designation.[91]

Amidst this push to exact India's compliance towards its "maximum pressure" policy, there was also a concurrent push of the Trump administration for US–India energy ties. Wherein, India's energy import from the US witnessed a meteoric rise. For instance, India began importing crude oil from the US "in 2017 as it looked to diversify its import basket beyond the OPEC nations. It bought 1.9 million tonnes (38,000 bpd) of crude oil from the US in 2017–18 and another 6.2 million tonnes (1,24,000 bpd) in 2018–19".[92] However, ahead of the end of the discussed exemption period, in the first half of the fiscal year 2019–20, the US "supplied 5.4 million tonnes of crude oil to India" to swiftly become India's sixth largest oil supplier.[93]

Apart from exacting Indian compliance on the Trump administration's "maximum pressure" policy against Iran, the US had also been amenable to some Indian considerations. Case in point, India's strategic investment in Iran, namely the development of Chabahar Port, was excluded from the purview of American sanctions, due to its potentialities towards the common US–India interest of encouraging landlocked Afghanistan's access to trade routes – and thereby strengthen the civilian dispensation in Kabul.

Support India's capacity-building

President Trump did not shy from touting defence exports as a US foreign policy priority, via instituting a precedent of the 'reverse revolving door', announcing the 'Buy American' policy to encourage arms sales, and construing defence trade as an incremental means to correct trade imbalances with partner nations.[94] Hence, with respect to India as well, the Trump administration not only oversaw the completion of deals finalised during the Obama administration (like the 2015 sale of AH-64E Apache helicopters) but also responded to Indian requests for particular defence platforms like the MH-60 Romeo Seahawk helicopter, which the Indian Navy had been looking to acquire for nearly a decade.[95]

In further encouraging US arms exports to India, the administration followed through on establishment recommendations to 'front-load' clearances ahead of Indian request for proposals. This encompassed the Trump administration wielding its executive authority, as in case of it overturning the Obama administration's freeze on India's acquisition of US-made unmanned systems – to effectively make India "the first non-treaty partner to be offered a MTCR Category-1 Unmanned Aerial System – the Sea Guardian UAS manufactured by General Atomics".[96]

Similarly, with Congressional directives on arms export designations, the Trump administration in its executive capacity proactively enacted them in case of India. For example, building on the Obama administration's Major Defense Partner status for India, the Trump administration accorded India with Strategic Trade Authorisation-1 (STA-1) status through a federal notification – making it only the third Asian country (after US allies Japan and South Korea) to have that status.[97]

Furthermore, the Trump administration also sought to cultivate a focus in US–India defence trade, towards India's maritime capabilities. The same seemed to stem from the rationale of facilitating a capable Indian maritime surveillance apparatus towards enhanced maritime domain awareness in the Indian Ocean region – which would then serve as a precursor to the US' longstanding aim of cultivating India's rise as a security provider in the region.

Beyond clearing the sale of the MH-60 Romeo Seahawk helicopters and India's additional procurement of the P-8 maritime surveillance aircraft, prominent examples under this focus included the Trump administration front-loading the clearance of the MK 45 5-inch/62-calibre anti-surface naval guns, MK 54 torpedoes, and AGM-84L Harpoon missiles – the latter two were aimed at enhancing India's P-8 aircraft fleet to conduct "anti-surface warfare missions in defense of critical sea lanes while enhancing interoperability with the United States and other allied forces".[98]

Thus, Trump's South Asia policy – which initially sought to break the policy inertia in Afghanistan – was eventually forced to recognise the imperatives of Pakistan's role in overseeing a resolution to America's longest war. This rendered the Trump administration's campaign of economic coercion towards Pakistan to have limited results, at times even at the risk of upending America's post–Cold War de-hyphenation of India and Pakistan. Moreover, with persistent trade tensions, the Trump administration even selectively adhered to the erstwhile Carter mantra, which had accorded a semblance of structure to the otherwise unformalised US–India strategic ties. However, in continuing the US' strategic courtship of India, the Trump administration's record on institutionalising strategic ties and mitigating India's westward apprehensions remains unprecedented.

Notes

1 David Shambaugh, "China Engages Asia: Reshaping the Regional Order", *International Security*, Volume 29, Issue 3 (Winter 2004/05), p. 95.

2 Michael Beckley, "The Myth of Entangling Alliances – Reassessing the Security Risks of U.S. Defense Pacts," *International Security*, Issue 39, Volume 4 (Spring 2015).

3 David Vine, "Where in the World Is the U.S. Military?", *Politico*, July/August 2015, last accessed 21 September 2020, www.politico.com/magazine/story/2015/06/us-military-bases-around-the-world-119321

4 DoD archive, "Unified Command Plan", The US Department of Defense, 28 October 2011, last accessed 21 September 2020, https://archive.defense.gov/ucc/

5 John F Kennedy, "Remarks of Senator John F. Kennedy, Conference on India and The United States, Washington, D.C., May 4, 1959", *John F. Kennedy Presidential Library and Museum*, 04 May 1959, last accessed 21 September 2020, www.

jfklibrary.org/archives/other-resources/john-f-kennedy-speeches/india-and-the-us-conference-washington-dc-19590504

6 *Quoted in* Guardian editorial, "The Guardian View on Kashmir: The World's Most Dangerous Place", *The Guardian*, 25 February 2019, last accessed 21 September 2020, www.theguardian.com/commentisfree/2019/feb/25/the-guardian-view-on-kashmir-the-worlds-most-dangerous-place

7 NSS 1996, "A National Security Strategy of Engagement and Enlargement", *National Security Strategy Archive*, February 1996, last accessed 21 September 2020, https://nssarchive.us/wp-content/uploads/2020/04/1996.pdf

8 K. Alan Kronstadt, "Pakistan–U.S. Relations", *Congressional Research Service*, 15 July 2019, p. 1, last accessed 21 September 2020, https://crsreports.congress.gov/product/pdf/IF/IF11270

9 Lalit K Jha, "US Slaps Sanctions on LeT, JeM Ahead of Obama Visit", *Outlook India*, 04 November 2010, last accessed 21 September 2020, www.outlookindia.com/newswire/story/us-slaps-sanctions-on-let-jem-ahead-of-obama-visit/700069

10 Donald Trump, "Remarks by President Trump on the Strategy in Afghanistan and South Asia", *The White House – Donald J Trump Administration*, 21 August 2017, last accessed 21 September 2020, www.whitehouse.gov/briefings-statements/remarks-president-trump-strategy-afghanistan-south-asia/

11 K. Alan Kronstadt, "Pakistan–U.S. Relations", *Congressional Research Service*, 15 July 2019, p. 1, last accessed 21 September 2020, https://crsreports.congress.gov/product/pdf/IF/IF11270

12 Donald Trump, "The United States Has Foolishly Given Pakistan More than 33 Billion Dollars in Aid over the Last 15 Years, and They Have Given Us Nothing but Lies & Deceit, Thinking of Our Leaders as Fools. They Give Safe Haven to the Terrorists We Hunt in Afghanistan, with Little Help. No More!", *Donald Trump – Twitter*, 01 January 2018, last accessed 21 September 2020, https://twitter.com/realDonaldTrump/status/947802588174577664?s=20

13 K. Alan Kronstadt, "Pakistan–U.S. Relations", *Congressional Research Service*, 15 July 2019, p. 2, last accessed 21 September 2020, https://crsreports.congress.gov/product/pdf/IF/IF11270

14 ANI report, "In Another Setback, US Cuts $440 Million Financial Aid to Pakistan", *India Today*, 17 August 2019, last accessed 21 September 2020, www.indiatoday.in/world/story/in-another-setback-us-cuts-440-million-financial-aid-to-pakistan-1581661–2019–08–17

15 Yashwant Raj, "Pakistan Lost $1.66 Billion in Trump's 2018 Aid Freeze", *The Hindustan Times*, 21 November 2018, last accessed 21 September 2020, www.hindustantimes.com/world-news/pakistan-lost-1-66-billion-in-trump-s-2018-aid-freeze/story-ybNB08pzRpzqgYlAgTx2sL.html

16 K. Alan Kronstadt, "Pakistan–U.S. Relations", *Congressional Research Service*, 15 July 2019, p. 2, last accessed 21 September 2020, https://crsreports.congress.gov/product/pdf/IF/IF11270

17 TOI report, "India, US Are One in Saying Pakistan Deserved to Be Demoted to Anti-terror Funding Group's 'Grey List'", *The Times of India*, 30 June 2018, last accessed 21 September 2020, https://timesofindia.indiatimes.com/india/india-us-are-one-in-saying-pakistan-deserved-to-be-demoted-to-anti-terror-funding-groups-grey-list/articleshow/64802488.cms?utm_source=contentofinterest&utm_medium=text&utm_campaign=cppst

18 Donald Trump, "Remarks by President Trump on the Strategy in Afghanistan and South Asia", *The White House – Donald J. Trump Administration*, 21 August 2017, last accessed 21 September 2020, www.whitehouse.gov/briefings-statements/remarks-president-trump-strategy-afghanistan-south-asia/

19 WSJ report, "America's Longest War: A Visual History of 18 Years in Afghanistan", *The Wall Street Journal*, 29 February 2020, last accessed 21 September 2020, www.wsj.com/articles/americas-longest-war-a-visual-history-of-18-years-in-afghanistan-11583010024
20 CRS report, "Afghanistan: Background and U.S. Policy: In Brief", *Congressional Research Service*, 25 June 2020, p. 1, last accessed 21 September 2020, https://fas.org/sgp/crs/row/R45122.pdf
21 David Welna, "Pentagon Questioned over Blackout on War Zone Troop Numbers", *NPR*, 03 July 2018, last accessed 21 September 2020, www.npr.org/2018/07/03/625544265/pentagon-questioned-over-war-zone-numbers-blackout
22 CRS report, "Afghanistan: Background and U.S. Policy: In Brief", *Congressional Research Service*, 25 June 2020, p. 8, last accessed 21 September 2020, https://fas.org/sgp/crs/row/R45122.pdf
23 CRS report, "Afghanistan: Background and U.S. Policy: In Brief", *Congressional Research Service*, 25 June 2020, p. 9, last accessed 21 September 2020, https://fas.org/sgp/crs/row/R45122.pdf
24 UNAMA report, "Afghanistan – Protection of Civilians in Armed Conflict 2019", *United Nations Assistance Mission in Afghanistan*, February 2020, p. 6, last accessed 21 September 2020, https://unama.unmissions.org/sites/default/files/afghanistan_protection_of_civilians_annual_report_2019_-_22_february.pdf
25 CRS report, "Afghanistan: Background and U.S. Policy: In Brief", *Congressional Research Service*, 25 June 2020, p. 12, last accessed 21 September 2020, https://fas.org/sgp/crs/row/R45122.pdf
26 *Quoted in* Jamie Crawford, "Top US Military Officer Says Taliban 'Are Not Losing'", *CNN*, 17 November 2018, last accessed 21 September 2020, https://edition.cnn.com/2018/11/17/politics/joseph-dunford-taliban-afghanistan/index.html
27 Donald Trump, "Remarks by President Trump and Prime Minister Khan of the Islamic Republic of Pakistan Before Bilateral Meeting", *The White House – Donald J. Trump Administration*, 22 July 2019, last accessed 21 September 2020, www.whitehouse.gov/briefings-statements/remarks-president-trump-prime-minister-khan-islamic-republic-pakistan-bilateral-meeting/
28 Mujib Mashal, "2 Weeks of U.S.–Taliban Talks End with 'Progress' but No Breakthrough", *The New York Times*, 12 March 2019, last accessed 21 September 2020, www.nytimes.com/2019/03/12/world/asia/afghanistan-us-taliban-talks.html
29 CRS report, "Afghanistan: Background and U.S. Policy: In Brief", *Congressional Research Service*, 25 June 2020, p. 11, last accessed 21 September 2020, https://fas.org/sgp/crs/row/R45122.pdf
30 Mujib Mashal, "2 Weeks of U.S.–Taliban Talks End with 'Progress' but No Breakthrough", *The New York Times*, 12 March 2019, last accessed 21 September 2020, www.nytimes.com/2019/03/12/world/asia/afghanistan-us-taliban-talks.html
31 Mujib Mashal, "2 Weeks of U.S.–Taliban Talks End with 'Progress' but No Breakthrough", *The New York Times*, 12 March 2019, last accessed 21 September 2020, www.nytimes.com/2019/03/12/world/asia/afghanistan-us-taliban-talks.html
32 BBC report, "Afghan Conflict: US and Taliban Sign Deal to End 18-year War", *BBC*, 29 February 2020, last accessed 21 September 2020, www.bbc.com/news/world-asia-51689443
33 DoD report, "Enhancing Security and Stability In Afghanistan", *The US Department of Defense*, December 2019, p. 24, p. 24, last accessed 21 September 2020, https://media.defense.gov/2020/Jan/23/2002238296/-1/-1/1/1225-REPORT-DECEMBER-2019.PDF
34 *Quoted in* IANS report, "Pompeo Meets Qureshi, Lauds Pak's Efforts for Afghan Peace", *Outlook India*, 18 January 2020, last accessed 21 September 2020,

www.outlookindia.com/newsscroll/pompeo-meets-qureshi-lauds-paks-efforts-for-afghan-peace-ld/1711390

35 Shakeel Qarar, "Wells Appreciates Pakistan's Efforts Against Illegal US Immigrants", *Dawn*, 20 January 2020, last accessed 21 September 2020, www.dawn.com/news/1529506

36 Shamil Shams, "US Approves Support for Pakistan's F-16s amid Afghanistan Outreach", *DW English*, 27 July 2019, last accessed 21 September 2020, www.dw.com/en/us-approves-support-for-pakistans-f-16s-amid-afghanistan-outreach/a-49766828

37 Newsweek report, "U.S. Reiterates Resumption of Military Training for Pakistan", *Newsweek Pakistan*, 04 January 2020, last accessed 21 September 2020, www.newsweekpakistan.com/u-s-reiterates-resumption-of-military-training-for-pakistan/

38 George Packer, "The Last Mission", *The New Yorker*, 21 September 2009, last accessed 21 September 2020, www.newyorker.com/magazine/2009/09/28/the-last-mission

39 USCENTCOM archives, "White Paper of the Interagency Policy Group's Report on U.S. Policy toward Afghanistan and Pakistan", *The US Central Command Archives*, 27 March 2009, last accessed 21 September 2020, www.centcom.mil/MEDIA/PRESS-RELEASES/Press-Release-View/Article/903723/white-paper-of-the-interagency-policy-groups-report-on-us-policy-toward-afghani/

40 Nicholas Kristof, "What Holbrooke Knew", *The New York Times*, 14 May 2011, last accessed 21 September 2020, www.nytimes.com/2011/05/15/opinion/15kristof.html

41 Nicholas Kristof, "What Holbrooke Knew", *The New York Times*, 14 May 2011, last accessed 21 September 2020, www.nytimes.com/2011/05/15/opinion/15kristof.html

42 Michael Hastings, "The Runaway General: The Profile That Brought down McChrystal", *Rolling Stone*, 22 June 2010, last accessed 21 September 2020, www.rollingstone.com/politics/politics-news/the-runaway-general-the-profile-that-brought-down-mcchrystal-192609/

43 Peter Feaver, "Bob Woodward Strikes Again! (McChrystal Assessment Edition)", *Foreign Policy*, 21 September 2009, last accessed 21 September 2020, https://foreignpolicy.com/2009/09/21/bob-woodward-strikes-again-mcchrystal-assessment-edition/

44 Peter Baker, "How Obama Came to Plan for 'Surge' in Afghanistan", *The New York Times*, 05 December 2009, last accessed 21 September 2020, www.nytimes.com/2009/12/06/world/asia/06reconstruct.html

45 Steven Erlanger, "Clinton Seeks Power to Lift India–Pakistan Sanctions", *The New York Times*, 14 July 1998, last accessed 21 September 2020, www.nytimes.com/1998/07/14/world/clinton-seeks-power-to-lift-india-pakistan-sanctions.html

46 *Quoted in* Mukund Padmanabhan, "How India Kept Kashmir out of U.S. Af-Pak Envoy's Brief", *The Hindu*, 21 May 2011, last accessed 21 September 2020, www.thehindu.com/news/the-india-cables/how-india-kept-kashmir-out-of-us-afpak-envoys-brief/article2035576.ece

47 Alyssa Ayres, "Trouble with the Facts When Trump Meets Imran Khan", *Council on Foreign Relations*, 22 July 2019, last accessed 21 September 2020, www.cfr.org/blog/trouble-facts-when-trump-meets-imran-khan?utm_medium=social_share&utm_source=tw

48 Donald Trump, "Remarks by President Trump and Prime Minister Khan of the Islamic Republic of Pakistan Before Bilateral Meeting", *The White House – Donald J. Trump Administration*, 22 July 2019, last accessed 21 September 2020, www.

whitehouse.gov/briefings-statements/remarks-president-trump-prime-minister-khan-islamic-republic-pakistan-bilateral-meeting/

49 Brad Sherman press release, "Statement by House India Caucus Co-chairs on Kashmir Dispute", *Website of Congressman Brad Sherman*, 22 July 2019, last accessed 21 September 2020, https://sherman.house.gov/media-center/press-releases/statement-by-house-india-caucus-co-chairs-on-kashmir-dispute

50 *Quoted in* Dipanjan Roy Chaudhury, "Donald Trump Repeats Kashmir Mediation Offer, India Says It's a Bilateral Issue", *The Economic Times*, 23 January 2020, last accessed 21 September 2020, https://economictimes.indiatimes.com/news/defence/donald-trump-repeats-kashmir-mediation-offer-india-says-its-a-bilateral-issue/articleshow/73529876.cms?utm_source=contentofinterest&utm_medium=text&utm_campaign=cppst

51 *Quoted in* Dipanjan Roy Chaudhury, "Donald Trump Repeats Kashmir Mediation Offer, India Says It's a Bilateral Issue", *The Economic Times*, 23 January 2020, last accessed 21 September 2020, https://economictimes.indiatimes.com/news/defence/donald-trump-repeats-kashmir-mediation-offerindia-says-its-a-bilateral-issue/articleshow/73529876.cms?utm_source=contentofinterest&utm_medium=text&utm_campaign=cppst

52 *Quoted in* Dipanjan Roy Chaudhury, "Donald Trump Repeats Kashmir Mediation Offer, India Says It's a Bilateral Issue", *The Economic Times*, 23 January 2020, last accessed 21 September 2020, https://economictimes.indiatimes.com/news/defence/donald-trump-repeats-kashmir-mediation-offer-india-says-its-a-bilateral-issue/articleshow/73529876.cms?utm_source=contentofinterest&utm_medium=text&utm_campaign=cppst

53 Business Today report, "Article 370 Revoked: US Says It's India's 'Internal Matter' in a Blow to Pakistan", *Business Today*, 08 August 2019, last accessed 21 September 2020, www.businesstoday.in/current/economy-politics/article-370-revoked-us-india-internal-matter-blow-pakistan/story/370888.html

54 James McBride, "The Risks of U.S. Steel and Aluminum Tariffs," *Council on Foreign Relations*, 08 March 2018, last accessed 21 September 2020, www.cfr.org/backgrounder/risks-us-steel-and-aluminum-tariffs

55 Jeff Smith, "Modi 2.0: Navigating Differences and Consolidating Gains in India–U.S. Relations", *The Heritage Foundation*, 05 August 2019, last accessed 21 September 2020, www.heritage.org/asia/report/modi-20-navigating-differences-and-consolidating-gains-india-us-relations

56 *Quoted in* Reuters report, "Donald Trump Says 'Getting Nothing' After India Reduces Tariffs on Motorcycles", *Zee News*, 27 February 2018, last accessed 21 September 2020, https://zeenews.india.com/international-business/donald-trump-says-getting-nothing-after-india-reduces-tariffs-on-motorcycles-2084725.html

57 USTR press release, "United States Will Terminate GSP Designation of India and Turkey", *The Office of the United States Trade Representative*, 04 March 2019, last accessed 21 September 2020, https://ustr.gov/about-us/policy-offices/press-office/press-releases/2019/march/united-states-will-terminate-gsp

58 Prabha Raghavan, "What Is GSP, and How Did India Gain from Being on US Trade Preference List?", *The Indian Express*, 05 March 2019, last accessed 21 September 2020, https://indianexpress.com/article/explained/explained-india-gained-from-being-on-us-trade-preference-list-what-happens-now-5611457/

59 Prabha Raghavan, "What Is GSP, and How Did India Gain from Being on US Trade Preference List?", *The Indian Express*, 05 March 2019, last accessed 21 September 2020, https://indianexpress.com/article/explained/explained-india-gained-from-being-on-us-trade-preference-list-what-happens-now-5611457/

60 *Quoted in* PTI report, "US Lawmaker Urges USTR to Delay GSP Decision till Indian Elections", *The Economic Times*, 30 March 2019, last accessed 21 September 2020, https://economictimes.indiatimes.com/news/economy/policy/us-lawmaker-urges-ustr-to-delay-gsp-decision-till-indian-elections/articleshow/68642969.cms?from=mdr

61 PTI report, "Trump Strips India of Preferential Trade Status, Gives Jaishankar His First Challenge", *The Print*, 01 June 2019, last accessed 21 September 2020, https://theprint.in/world/trump-strips-india-of-preferential-trade-status-gives-jaishankar-his-first-challenge/244276/

62 PTI report, "India Hits Back at US, Hikes Import Duty on 29 Products", *The Times of India*, 21 June 2018, last accessed 21 September 2020, https://timesofindia.indiatimes.com/business/india-business/india-hikes-import-duty-on-certain-us-products/articleshow/64678424.cms

63 USTR factsheet, "U.S.–India Bilateral Trade and Investment", *The Office of the United States Trade Representative*, 15 August 2019, last accessed 21 September 2020, https://ustr.gov/countries-regions/south-central-asia/india

64 USTR factsheet, "U.S.–India Bilateral Trade and Investment", *The Office of the United States Trade Representative*, 15 August 2019, last accessed 21 September 2020, https://ustr.gov/countries-regions/south-central-asia/india

65 Pooja Chaudhuri, "US Calling India a 'Developed Nation' Is No Cause for Celebration", *The Wire*, 17 February 2020, last accessed 21 September 2020, https://thewire.in/trade/us-india-trade-developed-nation

66 Alyssa Ayres, "U.S. Relations with India – Prepared Statement by Alyssa Ayres, Senior Fellow for India, Pakistan, and South Asia Council on Foreign Relations – Before the Committee on Foreign Relations, United States Senate, 2nd Session, 114th Congress", *Council on Foreign Relations*, 24 May 2015, last accessed 21 September 2020, www.cfr.org/content/publications/attachments/052416_Ayres_Testimony.pdf

67 Neha Dasgupta and Aditya Kalra, "Exclusive: U.S. Tells India It Is Mulling Caps on H-1B Visas to Deter Data Rules – Sources", *Reuters*, 19 June 2019, last accessed 21 September 2020, www.reuters.com/article/us-usa-trade-india-exclusive-idUSKCN1TK2LG

68 USTR factsheet, "FACT SHEET ON 2019 NATIONAL TRADE ESTIMATE: Key Barriers to Digital Trade", *The Office of the United States Trade Representative*, March 2019, last accessed 21 September 2020, https://ustr.gov/about-us/policy-offices/press-office/fact-sheets/2019/march/fact-sheet-2019-national-trade-estimate

69 Rounak Kumar Gunjan, "Govt Risks Trade Battle with US over Religious Opposition to Dairy Items from 'Non-Veg' Cows," *News18*, 06 March 2019, last accessed 21 September 2020, www.news18.com/news/india/blood-meal-cattle-the-non-negotiable-sticking-point-as-us-scraps-indias-preferential-trade-treatment-2057333.html

70 Biswajit Dhar and K.S. Chalapati Rao, "Why an India–US Free Trade Agreement Would Require New Delhi to Reorient Key Policies," *The Wire*, 23 September 2019, last accessed 21 September 2020, https://thewire.in/trade/why-an-india-us-free-trade-agreement-would-require-new-delhi-to-reorient-key-policies

71 Kashish Parpiani, "Understanding India–US Trade Tensions Beyond Trade Imbalances", *Observer Research Foundation*, 05 February 2020, p. 7, last accessed 21 September 2020, www.orfonline.org/research/understanding-india-us-trade-tensions-beyond-trade-imbalances-61077/

72 Adam Behsudi, "India Could Be Next on Trump's Investigate-and-tariff Hit List", *Politico*, 03 June 2019, last accessed 21 September 2020, www.politico.com/story/2019/06/03/trump-india-tariffs-1489947

73 Ana Swanson and Vindu Goel, "As Trump Visits India, a Trade Deal Remains Elusive", *The New York Times*, 25 February 2020, last accessed 21 September 2020, www.nytimes.com/2020/02/25/business/economy/trump-india-trade-deal.html
74 USTR factsheet, "U.S.–India Bilateral Trade and Investment", *The Office of the United States Trade Representative*, 15 August 2019, last accessed 21 September 2020, https://ustr.gov/countries-regions/south-central-asia/india
75 Alyssa Ayres, "A Field Guide to U.S.–India Trade Tensions", *Council on Foreign Relations*, 13 February 2020, last accessed 21 September 2020, www.cfr.org/article/field-guide-us-india-trade-tensions
76 Jeff Smith, "Modi 2.0: Navigating Differences and Consolidating Gains in India–U.S. Relations", *The Heritage Foundation*, 05 August 2019, p. 11, last accessed 21 September 2020, www.heritage.org/asia/report/modi-20-navigating-differences-and-consolidating-gains-india-us-relations
77 White House joint statement, "Joint Statement by President Obama and Prime Minister Singh of India", *The Donald Trump Administration – The White House Archives*, 08 November 2010, last accessed 21 September 2020, https://obamawhitehouse.archives.gov/the-press-office/2010/11/08/joint-statement-president-obama-and-prime-minister-singh-india
78 Business Standard report, "India and the United States Sign the Logistics Exchange Memorandum of Agreement (LEMOA)", *Business Standard*, 30 August 2016, last accessed 21 September 2020, www.business-standard.com/article/news-cm/india-and-the-united-states-sign-the-logistics-exchange-memorandum-of-agreement-lemoa-116083000261_1.html
79 State Department joint statement, "Joint Statement on the Inaugural U.S.–India 2+2 Ministerial Dialogue", *The US Department of State*, 06 September 2018, last accessed 21 September 2020, https://translations.state.gov/2018/09/06/joint-statement-on-the-inaugural-u-s-india-22-ministerial-dialogue/
80 Rajat Pandit, "India, US Agree to Expedite Work on BECA, Strengthen Defence Ties", *The Times of India*, 26 February 2020, last accessed 21 September 2020, https://timesofindia.indiatimes.com/india/india-us-agree-to-expedite-work-on-beca-strengthen-defence-ties/articleshow/74307875.cms
81 Shishir Gupta, "NSA Ajit Doval Heads for Washington to Get India–US 2+2 Dialogue Back on Track", *The Hindustan Times*, 28 March 2018, last accessed 21 September 2020, www.hindustantimes.com/india-news/nsa-ajit-doval-heads-for-washington-to-get-india-us-2-2-dialogue-back-on-track/story-ZlmTjaD50tlf5opbBBJFCM.html
82 Sridhar Kumaraswami, "2+2: India, US Talk Terror, Sign Pact on Military Tech", *Asian Age*, 20 December 2019, last accessed 21 September 2020, www.asianage.com/india/all-india/201219/22-india-us-talk-terror-sign-pact-on-military-tech.html
83 Sandeep Unnithan, "Exclusive: We Can Match China in the Indian Ocean Region, Says Navy Chief Sunil Lanba", *India Today*, 17 November 2018, last accessed 21 September 2020, www.indiatoday.in/magazine/interview/story/20181126-we-can-match-china-in-the-indian-ocean-region-admiral-sunil-lanba-1388904–2018–11–17
84 Prakash Karat, "Nationalism Made in USA", *Outlook India*, 29 January 2018, last accessed 21 September 2020, www.outlookindia.com/magazine/story/nationalism-made-in-usa/299717
85 Dinakar Peri, "Military Cooperation Group Dialogue Postponed", *The Hindu*, 08 March 2020, last accessed 21 September 2020, www.thehindu.com/news/national/military-cooperation-group-dialogue-postponed/article31016621.ece
86 Bilal Abdi, "Iraq Maintains Top Position as India's Crude Oil Supplier in 2018–2019", *The Economic Times*, 30 April 2019, last accessed 21 September 2020,

https://energy.economictimes.indiatimes.com/news/oil-and-gas/iraq-maintains-top-position-as-indias-crude-oil-supplier-in-2018–2019/69106973

87 Alex Ward, "The Neoconservative Fantasy at the Center of the Soleimani Killing", *Vox*, 03 January 2020, last accessed 21 September 2020, www.vox.com/2020/1/3/21048096/soleimani-strike-killing-neoconservatism-pompeo-trump

88 Clifford Krauss, "Trump Hit Iran with Oil Sanctions. So Far, They're Working", *The New York Times*, 19 September 2018, last accessed 21 September 2020, www.nytimes.com/2018/09/19/business/energy-environment/iran-oil-sanctions.html

89 Harsh Pant, "India's Iran Conundrum Is Back", *The Diplomat*, 26 April 2019, last accessed 21 September 2020, https://thediplomat.com/2019/04/indias-iran-conundrum-is-back/

90 Bilal Abdi, "Iraq Maintains Top Position as India's Crude Oil Supplier in 2018–2019", *The Economic Times*, 30 April 2019, last accessed 21 September 2020, https://energy.economictimes.indiatimes.com/news/oil-and-gas/iraq-maintains-top-position-as-indias-crude-oil-supplier-in-2018–2019/69106973

91 Shubajit Roy, "US to India: Helping You on Masood Azhar, So End Iran Oil Imports", *The Indian Express*, 24 April 2019, last accessed 21 September 2020, https://indianexpress.com/article/india/us-to-india-helping-you-on-masood-azhar-so-end-iran-oil-imports-5691315/

92 PTI report, "India's Import of US Oil Jumps 10-fold to 2,50,000 bpd", *The Economic Times*, 25 February 2020, last accessed 21 September 2020, https://economictimes.indiatimes.com/industry/energy/oil-gas/indias-import-of-us-oil-jumps-10-fold-to-250000-bpd/articleshow/74302108.cms?utm_source=contentofinterest&utm_medium=text&utm_campaign=cppst

93 PTI report, "India's Import of US Oil Jumps 10-fold to 2,50,000 bpd", *The Economic Times*, 25 February 2020, last accessed 21 September 2020, https://economictimes.indiatimes.com/industry/energy/oil-gas/indias-import-of-us-oil-jumps-10-fold-to-250000-bpd/articleshow/74302108.cms?utm_source=contentofinterest&utm_medium=text&utm_campaign=cppst

94 Kashish Parpiani, "India–US Defence Trade Continuity Under Trump", *Observer Research Foundation*, 02 July 2020, last accessed 21 September 2020, www.orfonline.org/research/india-us-defence-trade-continuity-under-trump-68919/#_edn58

95 Amrita Nayak Dutta, "Why India Wants to Buy the MH-60 'Romeo' Helicopters from the US", *The Print*, 18 November 2018, last accessed 21 September 2020, https://theprint.in/defence/why-india-wants-to-buy-the-mh-60-romeo-helicopters-from-the-us/151140/

96 State Department factsheet, "U.S. Security Cooperation with India", *Bureau of Political-Military Affairs – US State Department*, 04 June 2019, last accessed 21 September 2020, www.state.gov/u-s-security-cooperation-with-india/

97 Lalit K Jha, "India Third Asian Nation to Get STA-1 Status from US", *LiveMint*, 04 August 2018, last accessed 21 September 2020, www.livemint.com/Politics/GvUiFqQuVzLnBYuM5mwFuK/India-third-Asian-nation-to-get-STA1-status-from-US.html

98 DSCA press release, "India – AGM-84L Harpoon Air-Launched Block II Missiles", *Defense Security Cooperation Agency*, 13 April 2020, last accessed 21 September 2020, www.dsca.mil/major-arms-sales/india-agm-84l-harpoon-air-launched-block-ii-missiles

CONCLUSION

Donald Trump infused a sense of 'disruption' in US foreign policy by prioritising the renegotiation of America's trade relationships and undercutting multilateralism by withdrawing the US from the Paris Agreement, the UN Human Rights Council, etc. Furthermore, in case of specific regions, the Trump administration made a case for its 'disruptions' by vilifying the predecessor administration's approach in order to then break policy inertia and anticipate different outcomes. The chief example of this was the Trump administration's withdrawal from the Iran Nuclear Deal and increased alignment of Washington's policy aims in the Middle East with Sunni Arab nations.

In consideration of both these matrices, the Trump administration's foreign policy in the Indo-Pacific witnessed the least amount of 'disruption'. Certainly, the Trump administration's withdrawal from the Trans-Pacific Partnership (TPP), for instance, would be an untoward exception. However, there is evidence to support the claim that the Trump administration played a consequential role in crafting a credible long-term US policy in the Indo-Pacific.

Consider the announcement of the Indo-Pacific construct itself. The Trump administration built on the Barack Obama administration's Pivot/Rebalance to Asia policy by aligning America's political outlook with its strategic establishment's long-standing view on the interlinked destines of the Indian and Pacific oceans. Moreover, in addressing the lack of a clear region-wide US policy aim – other than the mere defence of the status quo against China's evolving heft – the Trump administration drew from regional partners' (chiefly, Japan's and Australia's) experiences to espouse the cause of a "Free and Open Indo-Pacific' (FOIP)". Thus, in direct contradiction with Trump's effort of "divorcing" values from US foreign policy,[1] his administration embraced a traditional neoclassical realist approach in the region to assert a liberal worldview that stands for the "respect for sovereignty and independence of all nations; peaceful resolution of disputes; free, fair, and reciprocal trade based on open investment, transparent agreements, and connectivity; and adherence to international law, including freedom of navigation and overflight".[2] Moreover, in purporting an American economic agenda for the region, the Trump administration in the case of the Indo-Pacific region dampened its 'America First' inclinations to embrace multilateralism and development aid. In addition, as the Obama administration built on the predecessor George W Bush administration's precedent of cultivating strategic partnerships in the region, the Trump administration, too, pursued the institutionalisation of strategic avenues in the cases of both nascent and traditional US partners.

Most importantly, in a testament to the lasting criticality of Trump's record in the Indo-Pacific region, there is little evidence to suggest that President Joe Biden will dispense the policy aim of preserving and advancing a Free and Open Indo-Pacific. In fact, ahead of the 2020 election, Democratic presidential nominee and former US Vice President Joe Biden even adopted the 'Indo-Pacific' moniker to outline his vision – with greater platitudes on America's "indispensable role" as the only distinguishable factor – for the region.[3]

In addition, in terms of the Trump administration's policies towards China, US partners in the Pacific, and the subregion of South Asia, distilling Trump's record in the Indo-Pacific region has yielded the following key takeaways.

On China, the Trump administration pursued a sustained confrontational posture. In settling the post–Cold War policy challenge of either engaging or confronting China, the Trump administration's approach unarguably emerged as a key shift in US foreign policy. Drawing on his 2016 election campaign, wherein Trump attributed the hollowing out of America's industrial base to China's rise, a decline in support for yesteryears' liberal internationalist idea of cultivating China's rise as a "responsible stakeholder"[4] was apparent. Specifically, in deriding the Washington establishment for being culpable in orchestrating America's "managed decline" in face of commercial interests on trade with China,[5] Trump rallied against the fallacies of the US' engagement–containment policy dissonance.

The Trump administration also followed through with its three-pronged approach of weaponising the US' trade deficit to wage an 18-month long trade war against China, increasing the significance of American Freedom of Navigation Operations (FONOPs), and stemming the tide against China's technological primacy with a global campaign against Chinese telecommunication giants. Furthermore, notwithstanding legitimate criticisms over the US mostly choosing to take a unilateral route on confronting Beijing, the Trump administration had no qualms in noting its intent to confront China. For instance, in its maiden National Security Strategy, the Trump administration announced the US to be shifting its national security priorities to addressing the re-emergence of a "great power competition" with China and Russia.[6] Moreover, given the fraying of American bipartisanship on foreign policy in recent times, Trump's confrontationist posture invited concerns over the global impact of heightened frictions between the world's two largest economies. Over time, however, owing to an emergent anti-China sentiment taking hold amongst the American electorate, Congressional Democrats only complemented Trump's approach to China by either grounding his administration's actions in legislative precedents or pushing the White House to also act on China's civil liberties record in Tibet, Xinjiang, and Hong Kong.

At the same time, with respect to the Pacific, the Trump administration focused on the further cultivation of strategic ties with traditional and nascent partners. With Japan and Australia, for instance, the Trump administration built on the predecessor administration's record to reiterate America's commitments. This also informed greater alignment between US–Japan, US–Australia or even US–Japan–Australia on issues such as demanding an international probe into the origins of the coronavirus pandemic, coalescing a common front against the foothold of Chinese telecommunications systems, expansion of the G-7 grouping, and revival of the Quadrilateral Security Dialogue. With South Korea and the Philippines, although the Trump administration witnessed a degree of downturn owing to rising Chinese influence, Trump furthered strategic ties with nascent partners like Vietnam, Indonesia, and Singapore.

Furthermore, despite withdrawing the US from the predecessor administration's cornerstone economic initiative, i.e. the TPP, the Trump administration presented an American proposition for transnational commerce and development aims in the region. Toward this end, with substantial assistance from bipartisan legislators on the Capitol Hill, the Trump administration established the US International Development Finance Corporation and reinvigorated the Export-Import Bank of the United States. In addition, the Trump administration also sought to showcase the high standards of the "economic pillar" of its Indo-Pacific strategy – vis-à-vis Chinese debt-trap diplomacy, with the announced Indo-Pacific Business Forum and the Blue Dot Network. Lastly, in dampening its opposition to multilateralism, a slew of development initiatives with partner nations were announced to cater to the region's infrastructure and energy needs.

In the subregion of South Asia, the Trump administration first sought to break the US' policy inertia by narrowing America's security aims in Afghanistan, empowering theatre commanders and the US Department of Defense with greater operational latitude to gain an on-ground initiative, and permitting US diplomats to directly engage with the Taliban. This approach also included increased international denunciation and curbing of US monetary assistance towards Pakistan, owing to its slow-walked efforts to aid American counterterrorism objectives in the Af-Pak region. From the standpoint of the Bush and Obama administrations' policy precedent of hyphenating Afghanistan and Pakistan – not only towards the US military effort in Afghanistan but also towards ensuring the de-hyphenation of India and Pakistan as a result – Trump's policy bore considerable continuity. However, as this approach yielded limited outcomes for the Trump administration's declared policy aim to "extricate" the US from its protracted military effort in Afghanistan,[7] a concurrent normalisation of US–Pakistan ties towards seeking the latter's assistance in pushing the Taliban to the negotiating table was evident. This *momentarily* led to a degree of hyphenation between India and Pakistan, especially in the context of the US offering to mediate between the two South Asian nations following India's abrogation of Article 370 in Kashmir.

However, those concerns gradually dissipated as it became clear that even the Trump administration's vocal apprehensions on India's trade barriers could hardly impact the consolidation of US–India strategic ties. Under Trump, this consolidation encompassed a push to US–India defence interoperability, clearing hurdles to US–India defence co-production and co-development, instituting consultative engagement platforms that reduce the relationship's dependency on chemistry between the top political leadership, and addressing India's westward apprehensions to oversee its integration under the US' eastward Indo-Pacific security matrix.

In summation, in the years to come, the Trump years will possibly be looked back upon as having borne the roots of a region-wide American approach in the Indo-Pacific – specifically in the context of ushering the emergence of an international consciousness (and possibly even a renewed US bipartisanship) on addressing the China challenge. With the ongoing coronavirus pandemic that is believed to have originated in China, consciousness on reducing dependencies on China has gained prominence. Thus not just the US, but with the European Union's increased attention to setting up safeguards against foreign investment by government-subsidised investors, for instance[8] – the Trump administration's concerns on China have resonated across the world. Regional players like India have also led the way on curbing the market-access for Chinese tech companies.[9] In addition, the Trump administration leaves behind the legacy of encouraging regional powers and US partners in particular to increasingly burden-share – not only in terms of security commitments but also with becoming increasingly vocal with their own policy positions on Chinese transgressions. In this aspect, following the Obama years' rigorous engagement with Southeast Asia that did little towards ASEAN's continued hedging on matters pertaining to the US–China rivalry, the Trump administration's seemingly offhanded approach led to ASEAN releasing its "Outlook on the Indo-Pacific", wherein its positions clearly aligned with the aims of a 'Free and Open Indo-Pacific'.[10] Lastly, the Trump administration elevated the Indian Ocean region's significance in America's security calculus via additional emphasis on continuing the institutionalisation of US–India strategic ties and aiding India's rise as a security provider in the Indian Ocean.

Most importantly, Trump pursuing considerable continuity on asserting America's commitment to the region – despite initial worries over the impact of his 'America First' impulses – went a long way in conveying to regional powers the insularity that the US accords to the Indo-Pacific region, despite grappling with declining domestic support for American internationalism.

Notes

1 Julian Borger, "Rex Tillerson: 'America First' Means Divorcing Our Policy from Our Values", *The Guardian*, 04 May 2017, last accessed 21 September 2020, www.theguardian.com/us-news/2017/may/03/rex-tillerson-america-first-speech-trump-policy

2 State Department report, "A Free and Open Indo-Pacific – Advancing a Shared Vision", *The US Department of State*, 04 November 2019, p. 6, last accessed 21 September 2020, www.state.gov/wp-content/uploads/2019/11/Free-and-Open-Indo-Pacific-4Nov2019.pdf

3 Biden Campaign policy paper, "Joe Biden's Agenda for the Indian American Community", *Biden-Harris 2020 Campaign Website*, n.d., last accessed 21 September 2020, https://joebiden.com/indian-americans/#

4 Robert Zoellick, "Whither China: From Membership to Responsibility? – Robert B. Zoellick, Deputy Secretary of State – Remarks to National Committee on U.S.–China Relations", *The United States Department of State – Archives*, 21 September 2005, last accessed on 21 September 2020, https://2001-2009.state.gov/s/d/former/zoellick/rem/53682.htm

5 PBS transcript, "Steve Bannon – First Interview with the Former Trump Adviser", *Frontline – Public Broadcasting Service*, 17 March 2019, last accessed 21 September 2020, www.pbs.org/wgbh/frontline/interview/steve-bannon-3/

6 NSS 2017, "National Security Strategy of the United States of America", *The Donald Trump Administration – The White House Archives*, December 2017, p. 27, last accessed 21 September 2020, www.whitehouse.gov/wp-content/uploads/2017/12/NSS-Final-12-18-2017-0905.pdf

7 Al Jazeera report, "Trump Wants Pakistan to Help 'Extricate' US from Afghanistan", *Al Jazeera*, 23 July 2019, last accessed 21 September 2020, www.aljazeera.com/news/2019/07/trump-pakistan-extricate-afghanistan-190722193747363.html

8 Jack Ewing, "Europe Takes Steps to Block Chinese Bargain Hunters", *The New York Times*, 17 June 2020, last accessed 21 September 2020, www.nytimes.com/2020/06/17/business/european-union-china-deals.html

9 Chidanand Rajghatta, "Taking Cue from India, US to Ban TikTok", *The Times of India*, 01 August 2020, last accessed 21 September 2020, https://timesofindia.indiatimes.com/world/us/taking-cue-from-india-us-to-ban-tiktok/articleshow/77306497.cms

10 ASEAN report, "ASEAN Outlook on the Indo-Pacific", *ASEAN*, 23 June 2019, last accessed 21 September 2020, https://asean.org/asean-outlook-indo-pacific/

EPILOGUE

Anticipating Biden's proposition for the Indo-Pacific

During the 2020 US presidential campaign, Joe Biden committed to pursuing a restorationist US foreign policy. Biden vowed to restore "our historic partnerships",[1] after four years of Donald Trump deriding the post–Cold War policy precedent on Washington underwriting the security of its allied and partner nations. However, the Indo-Pacific is being seen as an area, where there would be little divergence between Biden's propositions and Trump's record.[2] After all, the Indo-Pacific has witnessed the least 'disruption' at the hands of Trump's 'America First' approach.

Even as the Trump administration challenged precedents on maintaining an ever-expanding constellation of US allies around the world, the Indo-Pacific oversaw the bolstering of America's position. Underpinned with Trump's policy of sustained confrontation with China, his administration built on the predecessor Barack Obama administration's Pivot/Rebalance to Asia policy (discussed in Chapter 2). With the Indo-Pacific strategy – which belatedly accorded policy heft to the US security establishment's view on the interlinked destinies of the Indian and Pacific oceans – the Trump administration honed a constructive record with US partners in the region. This encompassed Trump's decision to uphold and follow through on Obama-era security commitments (discussed in Chapter 4), such as reaffirming the scope of the US–Japan alliance to cover contested territories and actualising goals for rotational US troop presence in Australia. Trump also furthered relations with nascent partners like Vietnam and India by increasing US support for their naval capacity-building and institutionalising new security commitments.

However, following the election, President-elect Biden seemed to have altered his plans for policy continuity in the Indo-Pacific. On the campaign trail, Biden often invoked the Trump administration's Indo-Pacific nomenclature. However, he eventually tweaked the normative aims of the US cultivating a "free and open" Indo-Pacific to "maintaining" a "secure and prosperous" Indo-Pacific.[3] Even with nominations for national security cabinet posts, the Biden campaign barely addressed the threat posed by China and at times even returned to using the 'Asia-Pacific' moniker.[4]

Questions over Biden's commitment

In its initial years, the Biden administration's priorities will understandably pertain to domestic concerns like stemming the tide against the coronavirus pandemic in the US; overseeing the distribution of vaccines across the length and breadth of the continental United States; introducing fiscal measures to address the downturn in the US economy; and, finally, addressing the unemployment crisis which at a point was considerably graver than the situation during the Great Recession.[5]

The foreign policy front, however, will also remain abound with pressing challenges – like managing outstanding nuclear proliferation matters (negotiations with North Korea, re-entering the Iran Nuclear Deal), reinvigorating international action on global governance challenges (like climate change), addressing the worldwide recession of liberal democracies (in face of trends favouring the allure of state-centred authoritarian models), and reasserting the post–Second World War dicta on a rules-based order (like denormalising territorial aggrandisement). Notably, all these challenges fundamentally hark back to the biggest geopolitical rivalry of the post–Cold War era. With China's central role in outstanding climate and proliferation matters, exporting the governance model of techno-authoritarianism, and flouting international norms in the South China Sea, US relations with Beijing will be at the centre of most foreign policy challenges awaiting Biden. Thus Biden's calibration of US policy towards China will be of utmost consequence.

Furthermore, in view of Trump's policy of sustained confrontation with China (discussed in Chapter 3) across multiple domains – trade, telecommunications, and maritime posture – any shift away from outright recognition of the strategic threat posed by China will only bear counterintuitive results: either in terms of further emboldening China to continue with its expansionist agenda (reflected in its 'Wolf-Warrior Diplomacy')[6] or diminished US credibility forcing regional countries to sue for peace with tacit accommodations of Beijing's heavy-handedness (as with the renewed talks for a China–ASEAN South China Sea Code of Conduct).[7]

Hence, for more reasons than one, the US–China dynamic has reached the proverbial 'end of the road' on balancing cooperation and confrontation. Not to mention, the fast solidifying US bipartisan consensus on confronting China at a time when rabid partisanship has consumed most avenues for cooperation between Republicans and Democrats. In possibly recognising this state of affairs, Biden's incoming National Security Advisor Jake Sullivan described China as "a serious strategic competitor".[8]

However, the discussed distancing from the Indo-Pacific nomenclature does spur consternation over Biden's commitment to the region and, by extension, to the continuation of a confrontational US posture against China. With Biden's announced national security cabinet being replete with veterans of the Obama administration, one cannot help but wonder if US activism in the Indo-Pacific will be the ill-fated casualty of a return to a conciliatory US policy towards China – much like how the Obama administration lost its tenacity to confront China (discussed in Chapter 1).

In seeking Chinese cooperation towards its plans for multilateral agreements on global governance issues, the Obama administration notably altered the six "key lines" informing the Pivot to Asia policy by including the prioritisation of a "stable and constructive relationship with China" and removing the emphasis on "forging a broad-based [US] military presence" in Asia. In a sign of a return to this Obama precedent on managing US ties with China, the Biden campaign has often coupled its pronouncements on the strategic threat posed by China, with the caveat that there "are areas where we will work with China when it's in our interests to do so on issues such as climate change".[9] In eyeing a window of opportunity, Beijing has even signalled its interest in Biden's restorationist agenda, by calling for the restoration of a "sensible approach" to US–China ties in view of "expanding converging interests by cooperation".[10]

While a serious policy argument can be made in favour of Biden trying to strike a balance in US relations with China, it is, however, a legitimate matter of concern if that would come at the cost of America's pivotal role in the Indo-Pacific. Under Obama for instance, a conciliatory policy towards China did translate into US ambivalence towards Chinese transgressions in the region. A key case in point is the US not holding Beijing to account on its declared assurance to not militarise the South China Sea. Similarly, concerns remain with Biden's commitment to once again have the US embrace multilateralism and lead the way on myriad global governance issues, which will invariably warrant Chinese cooperation. In the bargain for Chinese compliance with, say, the outstanding issue of Pyongyang's nuclear ambitions or expanding commitments under the Paris Climate Agreement, will Biden oversee a dampening of US commitments in the Indo-Pacific?

Biden will follow the Trump precedent

While Biden's shifting rhetoric may continue to pose questions over his commitment to the Indo-Pacific, some precedents are unlikely to be overturned.

To begin with, consider the Trump administration's decision to rechristen the US Pacific Command (PACOM) in Hawaii to the US Indo-Pacific Command. It is highly unlikely that Biden will reverse that decision, at the risk of undermining the strategic relevance of the Indian Ocean in the Indo-Pacific construct. In addition, Biden will most probably not overturn some administrative restructuring of the Trump years. For instance, the renaming of PACOM was followed by the Office of the Secretary of Defence having a policy department which clubs "allies and partners – Indo-Pacific maritime rim and island states – together in one component office, and China in another".[11]

In a sign of Biden continuing US focus on the region via its administrative structures, he has appointed former State Department official Kurt Campbell as the administration's "Indo-Pacific coordinator".[12] Most notably, in the first few months of the Biden administration, there has indeed been considerable US continuity in the Indo-Pacific. This has been apparent with the Biden administration's focus on expanding the Quad's purview beyond maritime issues and early high-level visits of Biden national security cabinet officials to Japan, South Korea, and India.

Furthermore, the Biden administration will most likely also face some political pressure to practise policy continuity in the Indo-Pacific. This may stem not only from the emergent bipartisan consensus on the matter but also from unequivocal pronouncements that the Trump administration made. This would include its National Security Strategy, which characterised the contention in the region as a "geopolitical competition between free and repressive visions of world order"[13] and not simply a matter of US–China geopolitical rivalry.

Similarly, in mid-January 2021, a week before the end of the Trump administration, the "United States Strategic Framework for the Indo-Pacific" was released. The political motivation to release the classified document 20 years ahead of time was apparent. In a statement announcing the declassification, Trump's National Security Advisor Robert O'Brien noted the rationale: "to communicate to the American people and to our allies and partners, the enduring commitment of the United States to keeping the Indo-Pacific region free and open long into the future".[14]

In addition, the document itself warned: "Loss of U.S. preeminence in the Indo-Pacific would weaken our ability to achieve U.S. interests globally".[15] Given the doubts over Biden's commitment to the region at the time, the warning seemed prophetic and directly relevant to Biden's restorationist foreign policy aim to "once more place America at the head of the table".[16]

Furthermore, over the past four years, as there has been an evident internationalisation of the Indo-Pacific construct, the Biden administration could face external pressure from subscribers to this construct. This pressure would stem from the Trump administration's precedent of bolstering participation of regional powers in the Indo-Pacific strategy (discussed in Chapter 4). Although this was often expressed with the Trump administration's harsh rhetoric on increased burden-sharing, the same left a precedent of increased participation from regional countries in terms of both contributing material resources *and* vocalising their positions on China's rise.

Over the past four years, this manifested in ways such as:[17] an evident surge in mini-lateral activity amongst regional powers (e.g. the forthcoming India–Japan–Australia supply chain initiative) and even with other external powers (e.g. the fast developing Paris–Delhi–Canberra partnership); regional powers like Australia leading the international initiative against the propositions of Chinese telecommunication giants and in calling for greater transparency over China's handling of the coronavirus pandemic; and, finally, with the region's preeminent multilateral forum, ASEAN shedding its hedging inclinations to underscore its "centrality" in the Indo-Pacific.

In underscoring US support for "allies' and partners' complementary approaches to regional engagement", O'Brien's statement thus noted this wide subscription to the Indo-Pacific construct.[18] It recounted complementary policies like, "Japan's *Free and Open Indo Pacific* concept, Australia's Indo-Pacific concept, India's *Security and Growth for All Regions* policy, the Republic of Korea's *New Southern Policy*, Taiwan's *New Southbound Policy*, and the Association of Southeast Asian Nations' (ASEAN) *Outlook on the Indo-Pacific*". On the construct's currency beyond the region, the statement noted that "many of these concepts and approaches are resonating globally, with countries such as France and Germany publishing their own policy frameworks for the Indo-Pacific".[19]

Going forward, prospective pushback from subscribers to the Indo-Pacific construct could increase, owing to China's actions to gain the initiative vis-à-vis regional countries, even as the world continues to grapple with the coronavirus pandemic.[20]

With Taiwan, China has stepped up its military intrusions in and across the Taiwan Strait.[21] With the infamous national security law in Hong Kong, Beijing has systematically sought to erode the policy of "one country, two systems".[22] With Australia, China has initiated a trade confrontation with the imposition of duties and has even sought to defame Canberra over its involvement in the US military effort in Afghanistan.[23] Similarly, China engaged in such actions with its neighbours even through 2020.[24] For instance, China and India engaged in a border skirmish in June 2020, which was deemed to be the first such instance between the two nations in over 40 years, and it claimed the lives of 20 Indian soldiers. China also stepped up its naval presence around Senkaku Island, which is claimed by Japan. China also officiated its claims in the South China Sea by announcing administrative jurisdictions over the Paracels Islands, Macclesfied Bank, and the Spratly Islands.

Finally, the Trump administration's economic policy in the Indo-Pacific seemingly aligns with the aims of Biden's intended foreign policy for the working class. The Biden's campaign's Build Back Better plan for jobs and economic recovery[25] could prove to be compatible with the Trump administration's record in the Indo-Pacific. Even as Trump embraced development aid and multilateral arrangements (discussed in Chapter 4) under the "economic pillar" of the Indo-Pacific strategy, his administration continued to seek expansion of market access for US businesses and thereby cultivate gains for the working class back home.

For instance, with its Asia EDGE initiative for leading the development of the region's energy infrastructure, the Trump administration evidently also sought "new export opportunities for American energy producers" under its policy of "unleashing American energy dominance".[26] With Trump having overseen the US' rise as the "world's largest energy producer" and a "net exporter of oil",[27] it will be hard for Biden to resist the inclination to further tap into the rising energy demands of the Indo-Pacific and in the bargain protect and cultivate jobs in the energy sector back home. Similarly, with respect to enhancing the manufacturing base in the US, it is unlikely that Biden will jettison the Trump administration's actions with the Indo-Pacific Business Forum and the reinvigoration of the US Export-Import Bank. According to State Department estimates, the latter alone cleared up exports worth $40 billion, which reportedly support nearly 230,000 jobs.[28]

Hence, even as questions remain over Biden's China policy and his commitment to the region, various administrative, political, regional, and economic precedents of the Trump era could provide the needed push to ensure US policy continuity.

Notes

1 Biden Campaign policy paper, "The Power of America's Example: The Biden Plan for Leading the Democratic World to Meet the Challenges of the 21st Century", *Biden-Harris 2020 Campaign Website*, n.d., last accessed 14 January 2021, https://joebiden.com/americanleadership/

2 Kashish Parpiani and Pratnashree Basu, "Biden and the Indo-Pacific – Will Regional Powers Shape America's Approach?", *Observer Research Foundation*, 10 January 2021, last accessed 14 January 2021, www.orfonline.org/expert-speak/biden-and-the-indo-pacific-will-regional-powers-shape-americas-approach/

3 PTI report, "Biden Emphasises on a Secure, Prosperous Indo-Pacific", *Deccan Herald*, 12 November 2020, last accessed 14 January 2021, www.deccanherald.com/international/biden-emphasises-on-a-secure-prosperous-indo-pacific-914587.html

4 Kashish Parpiani, "Lloyd Austin's Nomination Serves Biden's Political Calculations", *Observer Research Foundation*, 24 December 2020, last accessed 14 January 2021, www.orfonline.org/expert-speak/lloyd-austin-nomination-serves-biden-political-calculations/

5 Rakesh Kochhar, "Unemployment Rose Higher in Three Months of COVID-19 Than It Sid in Two Years of the Great Recession", *Pew Research Centre*, 11 June 2020, last accessed 14 January 2021, www.pewresearch.org/fact-tank/2020/06/11/unemployment-rose-higher-in-three-months-of-covid-19-than-it-did-in-two-years-of-the-great-recession/

6 Bloomberg report, "Welcome to Wolf-Warrior Diplomacy", *Bloomberg*, 03 December 2020, last accessed 14 January 2021, www.bloomberg.com/news/newsletters/2020-12-04/welcome-to-wolf-warrior-diplomacy

7 Laura Zhou, "South China Sea: China Asks ASEAN for Quick Resolution to Code of Conduct", *South China Morning Post*, 13 November 2020, last accessed 14 January 2021, www.scmp.com/news/china/diplomacy/article/3109706/south-china-sea-china-asks-asean-quick-resolution-code-conduct

8 *Quoted in* Fareed Zakaria, "On GPS: Biden's National Security Adviser on China", *CNN*, 03 January 2021, last accessed 14 January 2021, https://edition.cnn.com/videos/tv/2021/01/03/exp-gps-0103-jake-sullivan-china.cnn

9 Fareed Zakaria, “On GPS: Biden’s National Security Adviser on China”, *CNN*, 03 January 2021, last accessed 14 January 2021, https://edition.cnn.com/videos/tv/2021/01/03/exp-gps-0103-jake-sullivan-china.cnn

10 IANS report, “’Hope Next US Admin Under Biden Will Restore Normalcy to Bilateral Ties’: Chinese Foreign Minister Wang Yi”, *Times Now*, 02 January 2021, last accessed 14 January 2021, www.timesnownews.com/international/article/hope-next-us-admin-under-biden-will-restore-normalcy-to-bilateral-ties-chinese-foreign-minister-wang-yi/702032

11 Derek Grossman, “Biden Administration Could Benefit from Keeping an Indo-Pacific Focus”, *RAND*, 30 November 2020, last accessed 14 January 2021, www.rand.org/blog/2020/11/biden-administration-could-benefit-from-keeping-an.html

12 Josh Rogin, “Biden’s Pick for Top Asia Official Should Reassure Nervous Allies”, *The Washington Post*, 13 January 2021, last accessed 14 January 2021, www.washingtonpost.com/opinions/2021/01/13/kurt-campbell-biden-china-asia-nsc/

13 NSS 2017, “National Security Strategy of the United States of America”, *The Donald Trump Administration – The White House Archives*, December 2017, pp. 45–46, last accessed 14 January 2021, www.whitehouse.gov/wp- content/uploads/2017/12/NSS-Final-12–18–2017–0905.pdf

14 Robert O’Brien, “A Free and Open Indo-Pacific”, *The Donald Trump Administration – The White House Archives*, 05 January 2021, last accessed 14 January 2021, www.whitehouse.gov/wp-content/uploads/2021/01/OBrien-Expanded-Statement.pdf

15 NSC Declassified Document, “United States Strategic Framework for the Indo-Pacific”, *The Donald Trump Administration – The White House Archives*, 05 January 2021, last accessed 14 January 2021, www.whitehouse.gov/wp-content/uploads/2021/01/IPS-Final-Declass.pdf

16 Biden Campaign policy paper, “The Power of America’s Example: The Biden Plan for Leading the Democratic World to Meet the Challenges of the 21st Century”, *Biden-Harris 2020 Campaign Website*, n.d., last accessed 14 January 2021, https://joebiden.com/americanleadership/

17 Kashish Parpiani and Pratnashree Basu, “Biden and the Indo-Pacific – Will Regional Powers Shape America’s Approach?”, *Observer Research Foundation*, 10 January 2021, last accessed 14 January 2021, www.orfonline.org/expert-speak/biden-and-the-indo-pacific-will-regional-powers-shape-americas-approach/

18 Robert O’Brien, “A Free and Open Indo-Pacific”, *The Donald Trump Administration – The White House Archives*, 05 January 2021, last accessed 14 January 2021, www.whitehouse.gov/wp-content/uploads/2021/01/OBrien-Expanded-Statement.pdf

19 Robert O’Brien, “A Free and Open Indo-Pacific”, *The Donald Trump Administration – The White House Archives*, 05 January 2021, last accessed 14 January 2021, www.whitehouse.gov/wp-content/uploads/2021/01/OBrien-Expanded-Statement.pdf

20 Harsh Pant, “Joe Biden Is Behind the Curve on China’s Expansionism”, *Observer Research Foundation*, 17 December 2020, last accessed 14 January 2021, www.orfonline.org/research/joe-biden-behind-curve-china-expansionism/

21 John Xie, “China Is Increasing Taiwan Airspace Incursions”, *Voice of America*, 06 January 2021, last accessed 14 January 2021, www.voanews.com/east-asia-pacific/voa-news-china/china-increasing-taiwan-airspace-incursions

22 DW report, “Joshua Wong: ‘One Country, Two Systems’ Has Eroded to ‘One Country, One System’”, *Deutsche Welle*, 28 May 2020, last accessed 14 January 2021, www.dw.com/en/joshua-wong-one-country-two-systems-has-eroded-to-one-country-one-system/av-53601176

23 Ben Westcott, "China's Slur on Australian Soldiers Is Just the Latest Step in a Long Campaign by Beijing", *CNN*, 02 December 2020, last accessed 14 January 2021, https://edition.cnn.com/2020/12/01/asia/china-australia-afghanistan-us-intl-hnk/index.html

24 Kashish Parpiani and Dhriti Kamdar, "China's Expansionist Approach Is Raising Fresh Challenges for It, from the US to EU", *The Print*, 22 June 2020, last accessed 14 January 2021, https://theprint.in/opinion/chinas-expansionist-approach-is-raising-fresh-challenges-for-it-from-the-us-to-eu/446468/

25 Biden Campaign policy paper, "Build Back Better: Joe Biden's Jobs and Economic Recovery Plan for Working Families", *Biden-Harris 2020 Campaign Website*, n.d., last accessed 14 January 2021, https://joebiden.com/build-back-better/

26 WH factsheet, "President Donald J. Trump Is Unleashing American Energy Dominance", *The White House – Donald J. Trump Administration*, 14 May 2019, last accessed 14 January 2021, www.whitehouse.gov/ briefings-statements/president-donald-j-trump-unleashing-american-energy-dominance/

27 Patti Domm, "Biden Energy Plan May Be More Similar to Trump Policy Than Expected", *CNBC*, 02 November 2020, last accessed 14 January 2021, www.cnbc.com/2020/11/02/biden-energy-plan-may-be-more-similar-to-trump-policy-than-expected.html

28 State Department report, "A Free and Open Indo-Pacific – Advancing a Shared Vision", *The US Department of State*, 04 November 2019, p. 14, last accessed 14 January 2021, www.state.gov/wp-content/uploads/2019/11/Free-and-Open-Indo-Pacific-4Nov2019.pdf

INDEX

Made in the USA
Las Vegas, NV
29 October 2024

65f70ab9-268b-4049-bf92-fd52122f547eR01